"It's the job of a manager, not to light the fire of motivation, but to create an environment to let each person's personal spark of motivation blaze."

Frederick Herzberg

"You cannot mandate productivity. You must provide the tools to let people become their best."

Steve Jobs, CEO
Apple Computer
Pixar Animation Studios

RETURN ON INNOVATION™

TOOLS, TACTICS & TALENT™ FOR IMPLEMENTING WORLD-CLASS PRODUCT DEVELOPMENT

RETURN ON INNOVATION™

TOOLS, TACTICS & TALENT™ FOR IMPLEMENTING WORLD-CLASS PRODUCT DEVELOPMENT

BILL DRESSELHAUS, IDSA

Published by DRESSELHAUS DESIGN GROUP, Inc., Portland, Oregon USA

Manufactured in the United States of America.

ISBN: 0-9679209-0-6.
Front Cover Design: Michael Strickland, Strickland Design, Inc.
Graphic Design Support: Lisa Howell, John Schallberger, Jeff Lam, William W. Roberts, Feroshia Knight.
Production/Publishing Support: John Schallberger, Lisa Howell, Jeff Lam.
Editing: Jan Shannon, Lori Stephens, Robert Kalinowski, Chad Dresselhaus, Becky Dresselhaus, William W. Roberts.
Research Assistance: Lenka Dresselhaus, Chad Dresselhaus, Kathleen Anne.
Design, Writing, Marketing, and Publishing Consultants: William W. Roberts, Michael Strickland.
Art Direction, Book Design, Page Layout, Graphics/Diagram Development: Bill Dresselhaus.
Prototype Printing: Gary Walker, CTI Group, Portland, Oregon.
Prepress and Printing: Rob Goria, Paramount Graphics, Inc., Beaverton, Oregon.

DRESSELHAUS DESIGN GROUP, Inc.
8222 S.E. 162nd Avenue
Portland, Oregon 97236-4832
Studio/Fax: 503.760.1841
www.BillDresselhaus.com

This book is dedicated to my
wonderful wife Becky, who has, for
the past decade, patiently, lovingly and
persistently encouraged and
motivated me to write this book.

As with your favorite cookbook, *ROI: RETURN ON INNOVATION™* is designed for you to enter at the section that is most relevant to your immediate business needs—or to be reviewed and integrated into your company's infrastructure starting at page one. You are the chef, and these innovation recipes for success will impact your productivity for world-class product development.

EXPLORE - LEARN - INTEGRATE

BEGINNING STUFF	MANAGEMENT & FACILITATION	INNOVATION & DESIGN	MANUFACTURING & OPERATIONS	BUSINESS & MARKETING	ENDING STUFF

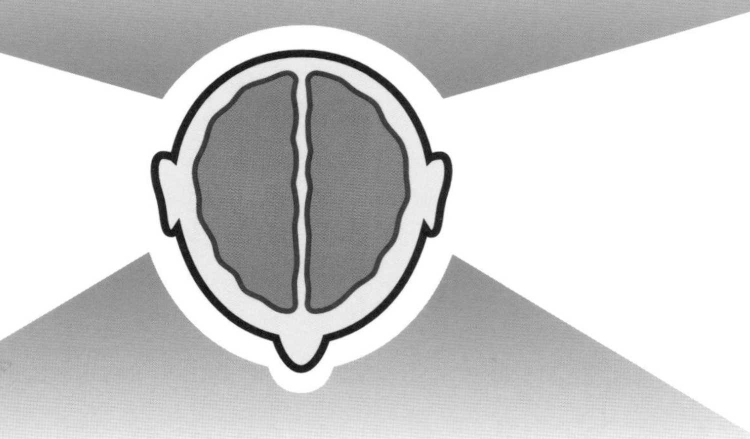

THE JOY OF INNOVATION

USING THIS BOOK

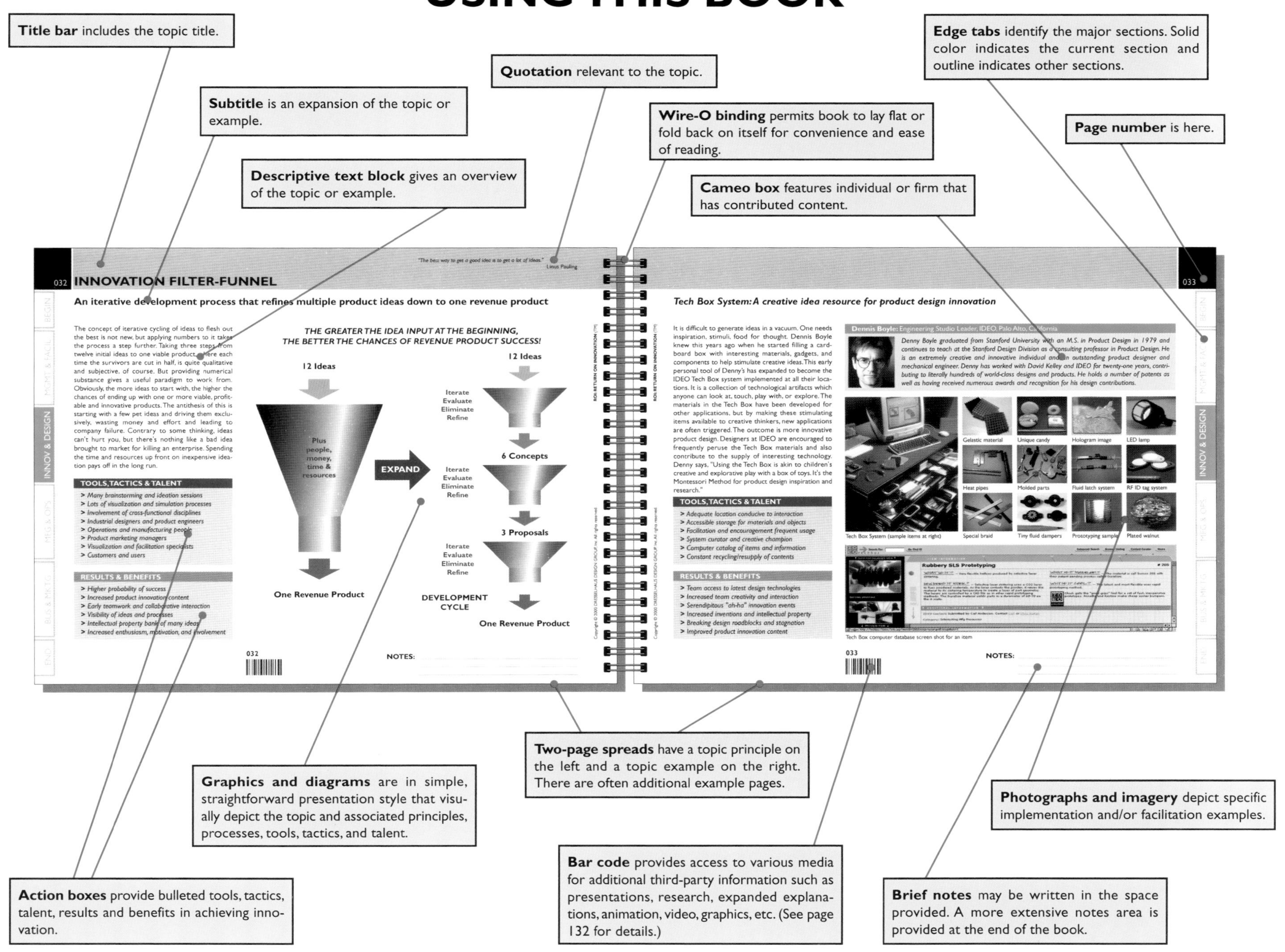

"Indeed, as much as one may resist it, we are gradually moving from a text-based to a visually oriented society."

Roger Cook in *Graphis Diagram 2*, 1996

"All good work is done in defiance of management."

Bob Woodward

BEGINNING STUFF

TABLE OF CONTENTS

BEGINNING STUFF
ix

MANAGEMENT & FACILITATION
001

INNOVATION & DESIGN
031

Table of Contents (continued)

ACKNOWLEDGMENTS

Many thanks to very many

Most books begin by observing that many people besides the author were integrally involved in bringing the book to the light of publishing day. While this may seem trite, it is particularly true of the book you hold in your hands. Indeed, a cursory glance at even a few spreads within these covers will make amply evident the truth of the observation as well as explaining the length of this acknowledgment section.

My special thanks go to Stanford University and its Design Division, first for letting me in and then for letting me out, with a degree no less. I have always been amazed that either event ever happened. I was fortunate that my mentor, Bob McKim, saw some spark of value in a frustrated and disillusioned chemical engineer who wanted to be an artist and designer. In addition, my thanks go to my other mentor at Stanford, Jim Adams, for letting me dangerously loose in his manufacturing course as a teaching assistant. I am forever grateful to these two men for their invaluable support, mentoring, and faith in me. They changed my life.

In addition to my Stanford mentors, several faculty at Art Center College of Design had a significant impact on my thoughts and skills in producing this book. Steve Diskin, Marty Smith, and my graduate faculty, Katherine Bennett, Steve Montgomery, David Lee, and Linda Shepard, all contributed training, inspiration, and ideas that focused and honed this book. Steve Diskin's contributions were especially significant, culminating in the fine foreword he wrote. Indeed, Steve's mentoring is the reason why this is a book rather than the video that I had initially planned for my graduate project. A special thanks to all of you at ACCD.

To Bill Roberts, who was and is weird enough to actually enjoy being involved at various stages of this project, my thanks. Since he had already been through the process of producing and self-publishing his own book, I asked him to be on board as I began the journey. At several critical junctures he lied to me about the complexities and difficulties of the process, thus encouraging me to proceed in the face of sometimes-daunting obstacles. His contributions included help with the initial concept, writing, reviewing, marketing, public relations, occasional graphic concepts, esoteric knowledge of arcane production issues, and general troubleshooting. Many thanks, Bill.

Next, thanks are due to the co-sponsors of this project for their valuable support, input, contacts, and encouragement. At SDRC, my special thanks go to Trudi Luebberst, Frank Kovacs, Garry Hearn, Tom Sigafoos, George Rendell, and Dan Meyer. I will always appreciate their ability to see that which didn't yet exist. At SGI, Steve Reckers, Denis Fastert, Susan Tellep, Judy Russell, and many others were integral in the sponsorship vision. At Alias|Wavefront, thanks are due to Thomas Heermann, Alexandra Walsh, Bill Buxton, Mark Sylvester, Lorraine McAlpine, Chris Cheung, Bill Coleman, and others. These people have the coolest toys and a great vision for virtual design. Thanks to all of you and your companies for your help and support.

A very special thank you goes to the folks at Acuity, Inc. Tom and Anne McKasson, who facilitated the introductions at SDRC, SGI, and A|W that started the sponsorship process, and Georgie Powers, Dave Crafton, Ace, Doug, Adrian, and the rest of the Acuity team, my thanks to all. Tom was the first to catch the vision for this book and has been a great encouragement throughout the project, especially during the many times when it seemed everything was going south on me.

My sincere gratitude goes to all of the collaborators for their great work and willingness to share it. Your insights, case studies, images, stories, and feedback were truly invaluable. Readers are encouraged to visit the Credits section of the book for all the details on these great people and companies where each collaborator is listed individually and by company.

A special set of supporters deserves separate recognition and gratitude: to Chad and Lenka Dresselhaus for their incredible research and data gathering work in several areas; to Jeff Lam, John Schallberger, Lisa Howell, and Feroshia Knight for their imaging, publishing, printing, and graphic design contributions; to Michael Strickland for a killer cover design; to Jan Shannon, Lori Stephens, and Bob Kalinowski for their excellent editing; to Kathleen Anne for her patient office management and diligent care of many routine tasks and details; and last but not least to John Eustermann, my lawyer, who writes bullet-proof contracts and gave me tons of great advice along the way that kept me out of lots of trouble.

Two printing companies were involved in the book production process: CTI Group (Gary Walker) for the many book prototypes along the way, and Paramount Graphics, Inc. (Rob Goria) for the final published product. In both instances, their expertise, patience, and diligence have helped bring this project to fruition. Many thanks!

Acknowledgments (continued)

Numerous photographers contributed images, and those whose names were known to me are appropriately credited. To those whose names were unknown, your work is nonetheless appreciated. Especially noted are Chris Skach for his great digital camera work and Steve Heller at Art Center College of Design for his invaluable layout advice. Thank you.

And, of course, supreme thanks and love to my wife, Becky, who has been a constant, solid source of encouragement, help, and wisdom as the book took its circuitous path toward completion.

Finally, I especially want to thank all the designers and engineers everywhere who get the job done every day in spite of tremendous obstacles and often poor management. You are the people who inspired me to do this book in hopes that it would help educate those who often don't understand why or how to get a return on their investment with innovation. Keep plugging away, keep developing great new products and satisfying customers, and keep modeling the creative processes that businesses and their management so desperately need today.

It was an honor for me to do this book with such great people and incredible talent. Any errors are mine, and any good is due to the contributions of all participants. Thanks all around. Now get out there and develop a great new product!

Bill Dresselhaus
Portland, Oregon
August 15, 2000

Much gratitude for funding, support, and encouragement

Structural Dynamics Research Corporation (SDRC)

Silicon Graphics, Inc. (SGI)

Alias|Wavefront Corporation

Acuity Incorporated

The corporate sponsors of this book have generously contributed funding, resources, and case study contacts. More importantly they have provided valuable advice, support, and motivation, which have helped to make the creation of this book a genuine pleasure. My sincere thanks to these world-class companies and their people.

Bill Dresselhaus

NOTES:

FOREWORD

Steve Diskin: Designer, Film Maker, Educator

Steve is an architect and industrial designer in Los Angeles. After his education at the Harvard University Graduate School of Design, he worked at Kenzo Tange + URTEC in Tokyo. His design of the HELIX clock (now in the permanent collection of the Cooper-Hewitt National Design Museum) and other products marked the beginning of his multi-disciplinary design career. Steve is a senior instructor in advanced product design and environmental design as well as coordinator of the graduate program in industrial design at Art Center College of Design in Pasadena, California. His work has been the subject of articles in the *New York Times*, *Los Angeles Times*, *Los Angeles Business Journal*, *Progressive Architecture*, and *Popular Science*. His multimedia videos won critical acclaim at the Industrial Designers Society of America conferences in 1994, 1995, and 1996. He divides his time between teaching, design consulting, and film making.

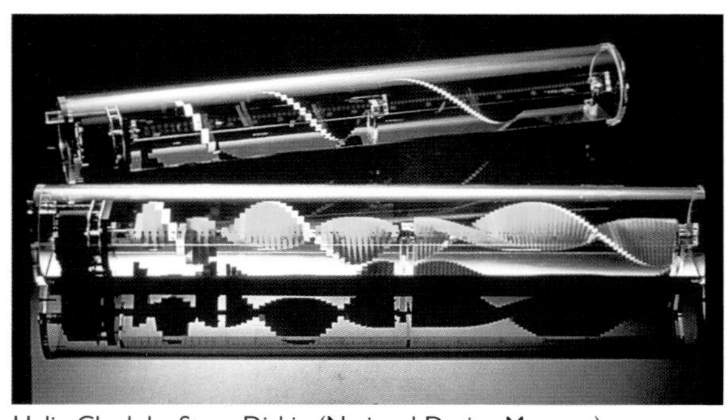

Helix Clock by Steve Diskin (National Design Museum)

Never before has the world of product development been so exciting. It's the beginning of a new millennium, after all, and we find ourselves in a world of technological and informational uproar, especially in design. What a crazy show it is—designers, engineers, marketing executives, vendors, and end-users all participating in the high drama of bringing technology, style, materials, and process together in a manufactured object. The play is complex and fascinating, sometimes difficult to comprehend, as we watch the action taking place. All too often we cannot see the players clearly at the back of the stage and we wonder why, for some strange reason, all these players are reciting their lines all at once!

In a way, it's amazing that anything ever gets produced. And it's also rather amazing that this drama is played out in corporate boardrooms and funky machine shops, in slick consulting offices and small design studios, in the quality control room and on the shipping dock, in stockrooms and testing labs all across the globe, and yet nobody seems to be watching. The drama of design has, for too long and far too often, been relegated to a small theater in the great scheme of life and commerce, out of view of the public eye. Nevertheless, it has been said that design is a process of narrative. Of story telling. It's really a variety of stories about how human beings apply their creativity to the creation of a new and better world. This story should be told to a wider audience!

There's a lot at stake here. As the cutting edge of technology, via rapid prototyping and computer graphics, makes the visualization (I should say *realization*) of a new product feasible in weeks, not months, so the need to understand product development more deeply and holistically becomes crucial. And if product development is so dramatically accelerating, how are we ever going to keep up with changing tastes, needs, and environmental issues, not to mention competition? We have to go back and look at the script to find out what we are really doing here.

I can think of no better guide to the drama of design than Bill Dresselhaus, a man who has seen firsthand the stories being told, both backstage and in front of the footlights. He has researched, designed, produced, and directed for most of his professional life, experiencing both the seriousness and the humor of the life cycle of products from start to finish. Here, in the pages of ***ROI: Return On Innovation***, Dresselhaus tells the stories of product development, introduces us to the players and their roles, and takes us behind the scenes for a revealing look at the process of design and mass production.

ROI is the program. Your seat is front row center. Join us now. The curtain is going up!

Steve Diskin, Coordinator
Industrial Design Graduate Program
Art Center College of Design
Pasadena, California
August 15, 2000

"*Weeks of research . . . Gut-level instinct . . . Same decision.*"
Anonymous

INTRODUCTION

Why another book on product development?

There are already plenty of good ones: Preston Smith's *Developing Products in Half the Time*, or Tom Peters' *The Circle of Innovation*, or Jerry Hirshberg's *The Creative Priority*, to name a few. What is needed is an innovation cookbook with images of the "food" and tantalizing, visualized "recipes" for getting there. Here are some of the reasons for this kind of book.

Executives and managers seldom deep-read. They don't have time. They skim or spot-read at best. Occasionally they drill down on a specific topic. Mostly, however, if they have any reading time at all, they review quickly for principles, images, and examples that are immediately obvious.

Strategies aren't everything. Corporate strategies and vision statements are relatively easy to come by. Most apply to just about any company: *Be the market leader in two years!; Consistently attain 40% profit margins!; Lead the industry with THE product of choice!* Etc, etc., etc. Who wouldn't want to attain these? Good stuff to strive for, but how does one get there? What's the recipe for success?

Tools, tactics, and talent. The key to strategic success is in knowing what to do at 8:00 on Monday morning that might actually begin to achieve those lofty corporate objectives. What should be done, what tools should be used, and who will do it? Printing the corporate vision statement on a little card for every employee to carry around and recite from memory just doesn't get it.

Business is about ideas. Jerry Hirshberg is right on: without a constant flow of innovative ideas to feed the corporate enterprise, any company will die no matter how good their quality, productivity, or business model. If the business of business IS ideas, as Jerry says, then it ought to be the number one priority of business to facilitate the tools, tactics, and talent that generate those new ideas.

It's facilitation instead of management. Management is about control. It's about budgets and finance, schedules and reviews, meetings and administration. The term "innovation management" may, in fact, be an oxymoron. Innovation and creativity must be facilitated, choreographed, and directed. It must be simmered, seasoned, stirred, and even allowed to boil over at times. But be careful of trying to *manage* it! The essence of innovation is getting out of the box. Control and management are the opposite.

This isn't rocket science. Being creative and innovative across the enterprise is not a complex task. It's pretty simple stuff that most any company can do. However, it does take a certain kind of attitude and commitment to new ways of approaching old problems. Priorities, paradigms, and processes must be modified and sometimes replaced. When you keep doing things the same way—expect to keep getting the same results.

Metrics can be deceptive. In a highly competitive market things are constantly in flux. By the time you have the resources and time to get data on a project, the world will have changed enough to make that data questionable. Gut-level intuition and experience about what works and what doesn't is often far better than gathering and analyzing numbers that may not have anything to do with predicting outcomes.

It's not ALL about money. People who think so are dangerous. Product development is no more all about money than life is all about sleep, food, air, and water. These are essential to sustaining life, just as money is to sustaining business. But life is not just about those elements, and neither is business just about money. Business, like life, is about ideas, solving problems, and creating something new.

The duality of business. Business has two mutually dependent and complementary sides, much like the dining room and the kitchen. One side—the side of results, infrastructure, business model, finance, shareholders, metrics, and revenue—is often the main focus, while the other side—the side of innovation, creativity, ideas, and product development—is frequently treated as a necessary evil, as if the former can be emphasized over the latter. It's just the opposite. Without continuously generating new ideas and revenue-generating products, the business side has no foundation.

It IS about people. Try putting a thousand dollars cash on a table and watch to see what happens. Obviously, nothing. It takes human effort—the human ingredient—to act upon that cash before there can be a return. Without the creative talents and skills of people, innovation and productivity would never happen, no matter how good the financing, business model, and infrastructure.

Looking through the lens. When contemporary business heroes such as George Lucas, Steven Spielberg, or Steve Jobs are in action, they are seldom in a boardroom presenting a business model or a profit/loss statement. They are usually behind the camera, in the editing room, in the design studio, or otherwise participating directly in the development process helping to produce revenue-generating products.

Introduction (continued)

Manage up, serve down. Taking care of the people who get the work done, rather than appeasing those removed from actual development work, is what's important in any industry. A successful manager will mentor, motivate, and support the people they are responsible for and will keep the management above them out of the way so the work can get done. David Kelley, founder of IDEO, says that true status is who comes up with the best ideas—not who's the oldest, who has the biggest title, or who's been at the company the longest.

Corporate culture wars. Within any corporate environment there are battles between different functional groups for turf and control. Frequently this is due to mutual misunderstanding of group goals, objectives, paradigms, operating modes, reward systems, and outcomes. Better understanding of those differences can result in a more harmonious, cooperative, and interactive development enterprise.

It's a visual world. The old saying that a picture is worth a thousand words is more relevant today than ever. Expanding that premise, a three-dimensional model can be worth a thousand pictures. Our culture is constantly bombarded with information and data. Time is at a premium and learning visually has become a necessity. Explaining and communicating ideas visually saves time and effort while improving communication quality, understanding, and speed.

The digital revolution. Besides being visual, the world is also becoming a digital one. In his new book, *Business @ the Speed of Thought*, Bill Gates admonishes everyone to get everything into digital format as quickly a possible, since digital manipulation and management is far easier and more comprehensive. This is now possible in product development, even with such manual processes as freehand sketching and model making. With 2D and 3D scanners and a variety of exciting new technologies, nearly everything can be ultimately digital.

Bottom line. This book has been designed more like a product than like a publication. It is a communication product—designed, developed, and produced to meet a specific set of user needs. Its development has followed the principles and processes described within the book itself. Like any other product, this one will evolve, be refined, and mature into new and unexpected forms over time. I look forward to my readers participating and collaborating in this process. My hope is that this book will successfully communicate and demonstrate the innovation principles, processes, and people that can empower many to have fun, make money, and change the world!

The way things work (or should!)

On the next page is a diagram depicting the six generic phases of the product development process. This will come as a surprise to some people, who may think that six phases is about four too many. Some think products go through a quick design phase based on the first idea and then immediately to full production. Others might provide a short interim phase for manufacturing setup. Others, however, think that products, like babies, appear magically under toadstools. There are at least two problems with these simpler models. First, each of the six phases requires its own time and resources. Ignoring a phase misses the output that comes from it, impoverishing the next phase as well as the end product. Second, each stage needs proactive management, unlikely to happen if management is unaware of the existence or purpose of that phase. In either case, the result will be a misconceived, malnourished new product, unlikely to survive to adulthood.

TOOLS, TACTICS & TALENT

> *Broad understanding of the development process*
> *Management commitment to support all phases*
> *Right tools, tactics, and talent for each phase*
> *Proper inter-phase pauses, reviews, and checks*
> *Broad collaboration and dedication of entire team*
> *Intense early comprehensive concept resolutions*
> *Initial planning involvement of entire team*

RESULTS & BENEFITS

> *Best-practices process development*
> *Optimized project schedules and performance*
> *Minimized time-to-market of products*
> *Innovative, thorough, and robust product designs*
> *Maximized development projects ROI*
> *Strategic competitive market advantage*

The people below are talented and experienced industrial designers and product engineers and are well practiced in the art of product design and development. Each of them contributed to the process chart on the following page in some way. The chart should be used as a general guide to laying out a product development project. It can be applied to both productization and technology development but is more suited to the former.

Larry Spreckelmeier: Manager, Industrial Design Group, Ethicon Endo-Surgery, Inc., Cincinnati, Ohio

Larry originally created and staffed the Ethicon Endo-Surgery Rapid Prototype Center and the Industrial Design Group that supports overall development and operational functionality of Ethicon's medical equipment. With over thirty years of product development experience, Larry has previously managed projects for the appliance and toy industries and has consulted and managed aircraft hangar design and construction for the U.S. Air Force. Larry has a B.S. in industrial design and holds twelve medical design patents.

Terry Joehnk: Principal, Contract Design, Portland, Oregon

Terry has over ten years of experience in the field of industrial design and product development. He has a B.S. degree in industrial design from Arizona State University and is principal of Contract Design, a product development resource he founded to serve clients in the high technology product arena including computer furniture, computer accessories, communications products, and exhibit design. Terry serves a number of clients with product design and development support, including Anthro Technology Furniture and Command Communications, Inc.

Robert Williams: Senior Product Design Engineer, InFocus Corporation, Wilsonville, Oregon

Bob has eleven years of experience as a design engineer. While at InFocus, Bob has led the product design for eight projects, including four projectors, investigated R&D issues such as the alignment of small pixel LCDs and lamp chemistry and cooling, and has designed various simple and complex parts using sophisticated MCAD software. He has worked in environments that utilized small, focused teams to conceptualize and prototype designs and large teams that emphasized theoretical analysis to visualize and develop products. Bob received a B.S. in mechanical engineering and has eight years of MCAD solid modeling experience and six years designing opto-mechanical systems.

NOTES:

The Product Development Process

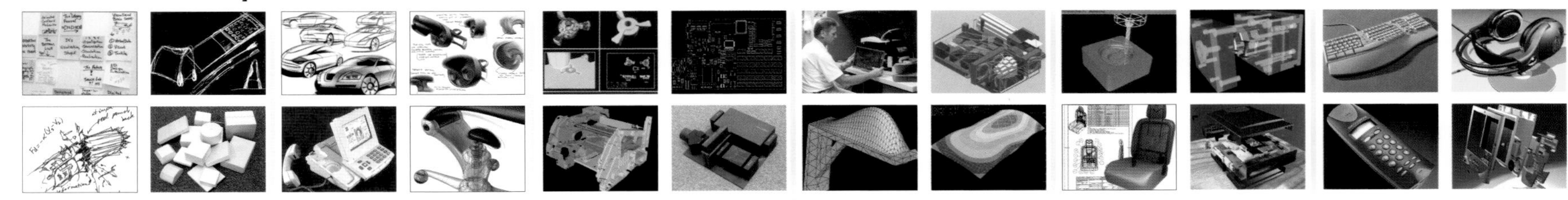

Phase 1 CONCEPT DEVELOPMENT	Phase 2 DESIGN DEVELOPMENT	Phase 3 ENGINEERING DEVELOPMENT	Phase 4 VALIDATION DEVELOPMENT	Phase 5 MANUFACTURING DEVELOPMENT	Phase 6 PRODUCTION DEVELOPMENT
TOOLS (*Project Mgt. Software*) > Hardware, software, materials, toys, tools, space > Interviews, observations, and focus groups > 2D and 3D physical and virtual visualization tools > World Wide Web and Internet access > Dedicated war room or interaction space > Facilities for interaction and presentation	**TOOLS** > 2D and 3D sketching tools and materials > CAID and MCAD conceptual/design software > RPM and 3D physical/virtual modeling tools > Simulation/technology laboratory and facilities > Dedicated war room or interaction space/tools > Workspaces fostering creativity and innovation	**TOOLS** > 2D and 3D sketching tools and materials > MCAD and CAE development tools > RPM and modeling resources and services > Physical and virtual analysis and test tools > Analysis and testing resources and facilities > ECAD tools and equipment	**TOOLS** > Virtual test, analysis, and validation tools > Test, analysis, and validation equipment > Test and validation laboratories/facilities > Machining/fabrication tools and shop > RPM resources and suppliers > CAD/CAE design tools	**TOOLS** > Assembly tools and equipment > CAM, simulation, and manufacturing tools > Tooling and fabrication suppliers > Secondary operations suppliers > Machine, prototype, and model shop/tools > Dimensional analysis and inspection tools/facilities	**TOOLS** > Facilities, process, and operations planning tools > Manufacturing and assembly line planning tools > QA/QC inspection tools and facilities > Product and component inventory control tools > Production validation and testing tools/facilities > Media and graphic development tools
TACTICS (*Project Management*) > Research and information gathering > 2D and 3D idea sketching and simulation > Problem definition and need finding > Studies of competition and previous products > Brainstorming and idea-generating sessions > Risk, technology, usability, and market assessments	**TACTICS** > 2D and 3D product concept simulations > Form, features, and appearance exploration > Product architecture and configuration design > Technology bread-boarding and validation > Usability, interaction, and marketing validation > Concept refinement and engineering definition	**TACTICS** > 2D and 3D engineering concept sketching > Electromechanical/technology engineering design > Detailed part and assembly design > Preliminary RPM of all parts and assemblies > Manufacturing, component, and tooling analysis > Preliminary virtual/physical validation testing	**TACTICS** > Prototype fabrication and assembly > Beta test program/process > Test and validation processes/procedures > Failure mode and environmental testing > User and usability testing > Design database refinement/revision	**TACTICS** > Supplier negotiation, qualification, and setup > Tooling and fabrication design/build > Secondary operations design/build > First article inspection and qualification > Detail design and component refinement/revision > Pre-production unit testing, analysis, and planning	**TACTICS** > Dimensional inspection and analysis process > Quality assessment and control processes > Revalidation test program with production units > Final agency certification/compliance process > Facilities, process, and assembly line planning > Advertising, promotion, and package design
TALENT (*Project Manager*) > Visualizers, researchers, interviewers, facilitators > Industrial designers and product engineers > Manufacturing, tooling, and production engineers > Human factors and ergonomic specialists > Sales, business, and marketing specialists > Expert users and potential customers	**TALENT** > Visualizers and simulation facilitators > Industrial designers and product engineers > Manufacturing and marketing specialists > Human factors and ergonomic specialists > Intellectual property and patent lawyers > Expert users and potential customers	**TALENT** > Modeling and simulation specialists > Industrial designers and product engineers > Technologists and physical scientists > Manufacturing and marketing specialists > RPM and analysis/test specialists > Manufacturing and tooling engineers	**TALENT** > Test and validation engineers/specialists > Component and quality engineers/specialists > Tooling and manufacturing engineers > Design and engineering team > Digital documentation specialists > Fabricators/modelmakers/technicians	**TALENT** > Manufacturing, fabrication, and tooling engineers > New Product Introduction (NPI) team > Purchasing and procurement specialists > Quality control and inspection specialists > Documentation and data management specialists > Product design and development team	**TALENT** > Manufacturing and production engineers > Quality and component engineers > Technical writers and graphic designers > Customer service and maintenance specialists > Marketing, advertising, and sales personnel > Purchasing and procurement specialists
DELIVERABLES (*Project Plan*) > Preliminary product proposal with project plan > Preliminary Marketing Requirements Document - Business plan and market strategy - Roadmap and platform plan > Preliminary Engineering Requirements Document - Technology and competition overview	**DELIVERABLES** > Revised/updated MRD, ERD, and product proposal > Industrial design and usability specification > Industrial design appearance model > Mechanical package design specification > Product architecture configuration prototype > Technology validation prototype	**DELIVERABLES** > Complete MCAD/CAE virtual database > Complete physical/mechanical prototype model > Revised final mechanical design specification > Comprehensive prototype validation test plan > Preliminary tooling and fabrication plan > Engineering design review and report	**DELIVERABLES** > Digital documentation package > Tooling plan review and release for tooling > Test, validation, and quality report > Final design review and report > Fully functional prototype(s) > Tooling and fabrication plan	**DELIVERABLES** > Final Bill of Materials (BOM) > Final documentation package > Pre-pilot report and evaluation > NPI and manufacturing/production plan > Supplier agreements and process plans > Production costing and financial analysis	**DELIVERABLES** > Continuous mass-produced product > Product packaging design and production > User and service manuals > Service and maintenance program > Advertising and promotion program > Comprehensive production and quality plan

DESIGN

ENGINEERING

MANUFACTURING

BUSINESS

DESIGN	ENGINEERING	MANUFACTURING	BUSINESS
> Industrial designers and interaction designers > Visualizers, graphic designers, and animators > Product designers and product engineers > Model makers and prototypers > Human factors and ergonomic specialists > CAID/CAD implementers and simulators	> Mechanical, software, and electronic engineers > Regulatory and component engineers > Manufacturing, test, and quality engineers > "Physics" engineers (optical, thermal, audio, etc.) > Electromechanical technicians and specialists > CAE and MCAD implementers and specialists	> Tooling engineers and fabrication specialists > Manufacturing, quality, and component engineers > Production managers and coordinators > Purchasing, procurement, and supplier specialists > Process engineers and production planners > CAM and data management software specialists	> Business modelers, simulators, and planners > Accounting and computational specialists > Marketing communications and sales specialists > Forecasting and finance specialists > Advertising, promotion, and branding specialists > Business and computational software specialists

This chart represents a general schematic product development process that will vary in detail between projects and products. It is intended to be a basic guide to the primary phases and tools, tactics, and talent needed to get a product successfully to market with innovation, thoroughness, and quality. This process is both cyclic (dashed arrows) and linear (large arrow). Thus, a balance between continuous iteration and simply getting done must be accomplished.

" . . . a traditional bureaucratic structure, with its need for predictability, linear logic, conformance to accepted norms, and the dictates of the most recent 'long range' vision statement, is a nearly perfect idea-killing machine."

Jerry Hirshberg in *The Creative Priority*

"Everybody wants results, but nobody is willing to do what it takes to get them!"

Clint Eastwood as Lt. Callahan

"We have met the enemy and it is us!"

Pogo

MANAGEMENT & FACILITATION

> *"[Leo] wanted to be more than a shopkeeper . . . a problem-solver at heart, he wanted a challenge."*
>
> Richard Smith

One of the most prolific and innovative product development geniuses of all time

Every Silicon Valley enthusiast knows the story of Bill Hewlett and Dave Packard, two young Stanford University engineers who worked in their garage in Palo Alto, California, designing and building instruments for Disney and the U.S. Government in 1944. This small, early high-tech endeavor resulted in the corporate giant, Hewlett-Packard. A similar story, however, also with a "garage" and humble beginnings, is perhaps even more remarkable. Clarence Leo Fender was an electronics repairman who began his business by soliciting door-to-door for radio repair. Leo eventually designed and developed one of the most famous and popular electric guitars in history, the Stratocaster. The lasting revenue value of the instrument he designed in the 1950s, during the same era as Bill and Dave, is remarkable. Although the first H-P products are now obsolete, Leo's "Strat" is still being sold and used in virtually the same form as the original design.

LEO FENDER: The Man

- *Born in a barn near Anaheim, California*
- *Lost eyesight in one eye as a child*
- *No college degree (only two years)*
- *Started as a radio electronics repairman*
- *Never learned to play a guitar*
- *Insatiable curiosity about how things should work*

LEO'S LEGACY: Great Process

- *Immersed in sensing and meeting customer needs*
- *Mutual engagement and collaboration with users*
- *Effective integration of varied product innovations*
- *Consideration of all enterprise segments*
- *Designed for superior user experience*
- *Dogged determination to get it right*

Clarence Leo Fender (deceased): Founder, Fender Guitar Company, Fullerton, California

While Leo Fender studied accounting in junior college, he mastered electronics on his own. Despite obstacles, Leo designed and built, among many products, several revolutionary guitars. These included the Telecaster, the first successfully mass-produced solid-body electric guitar that maintained a clean, amplified sound. Leo began his musical instrument empire in a radio repair shop in 1946 and subsequently sold his Fender Guitar Company to CBS for $13 million in 1965. He continued innovating, designing, and building new products until his death. His passion and his products live on today.

The Stratocaster U.S. patent

Early production model Stratocaster: S/N 0100

The original Fullerton, California, "garage" where Leo started

Leo Fender at work in his shop

NOTES: ______

World-Changing Design: A product design still making a profit, virtually unchanged since 1954

Designing and developing a successful and profitable product is a rewarding accomplishment in itself. But to have a design that remains so essentially unchanged for over forty years is phenomenal. The Fender Stratocaster solid-body electric guitar, designed in 1954 by Leo Fender, is arguably the most successful, most copied, and most popular musical instrument of its kind. Today's Stratocasters are essentially the same basic design as they were in 1957 and continue to make a profit for the Fender Musical Instruments Corporation, which produces new models and variations each year. It is an incredibly comfortable and versatile guitar that has been used at one time or another by virtually every great rock musician. It has literally changed the world of rock music (AND helped start it!) and has affected our culture forever.

REVOLUTIONARY FEATURES

> *Bolt-on replaceable neck*
> *Straight string pull to reduce stress*
> *Contoured body for optimum comfort*
> *Tremolo bridge for tonal variations*
> *Simple manufacturing processes*
> *Flat body/neck/headstock to reduce breakage*
> *Modular electronics for flexibility*
> *User-friendly and ergonomic features*

RESULTS & BENEFITS

> *Long-term success and profitability*
> *International brand recognition and respect*
> *Outstanding customer/user satisfaction*
> *Envied, imitated and copied by competition*
> *Low-cost production due to simple design*
> *Ease of repair, maintenance, and modification*

Fender Musical Instruments Corporation: Scottsdale, Arizona, and worldwide

Legendary rock, pop, and country-western musicians, including Jimi Hendrix, Eric Clapton, Buddy Guy, and Stevie Ray Vaughn, have played Fender Stratocaster guitars for decades. Fender Musical Instruments Corporation (FMIC) is arguably the world leader in the production of solid-body guitars and guitar amplifiers. FMIC has established highly specialized manufacturing and production operations in several countries. It also manufactures and distributes a full range of its products to wholesale and retail outlets throughout the world.

Latest Fender Stratocaster versions

Pure design simplicity: body, neck, electronics, and hardware

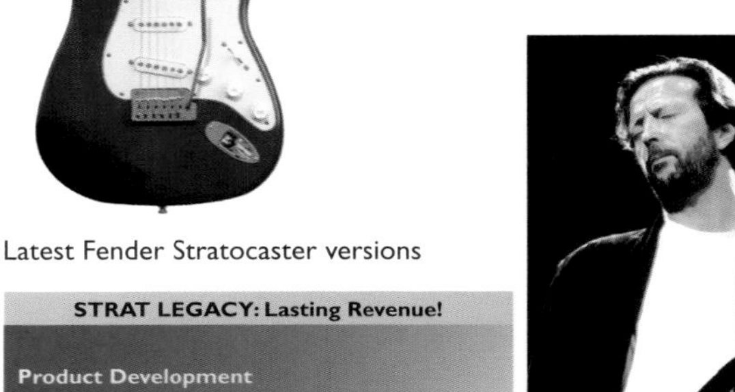

Revenue-generating design for 45 years

Eric Clapton

Jimi Hendrix

Stevie Ray Vaughn

NOTES:

"Diagrams are an alternative language which translates complexity into clarity at a time when people are skimming more and reading less."
Roger Cook in *Graphis Diagram 2*, 1996

VERBALLY BASED INTERACTION

The inadequacy of using only words and numbers to describe physical/visual ideas

The most common forms of communication in a product development environment are verbal and alpha-numeric. Visuals and graphics are being used more, but these are still too small a part of the typical materials. Poor or inadequate communication leads to incorrect mental models and ultimately to potentially inappropriate actions and reactions. What an audience perceives as reality is what they will go away with and act upon. Incorrect perception will lead to bad actions, bad planning, bad decisions, and subsequently bad results. When communicating ideas about visual and physical entities, an audience must translate verbal descriptions into their best mental models. Typically, everyone will have varied perceptions based on their interpretation. It is not surprising then, with vastly different mental models of what has been presented, that these individuals go away and act upon them inappropriately. The results are generally unfortunate.

Tom Sigafoos: SDRC Education Consortium Manager, SDRC, Cincinnati, Ohio

Tom has been instrumental in establishing relationships with over 250 schools in the U.S. and Canada. Most recently, he collaborated with McGraw-Hill, who published the SDRC Student Edition, a cost-effective tool for students in engineering. Tom has a B.A. and an M.A. and has served on many university steering committees. He is also a distinguished speaker who has been recognized as an authority on educational standards and principles. Tom has a special interest in fostering good communication and mutual understanding of expectations in management/staff relationships.

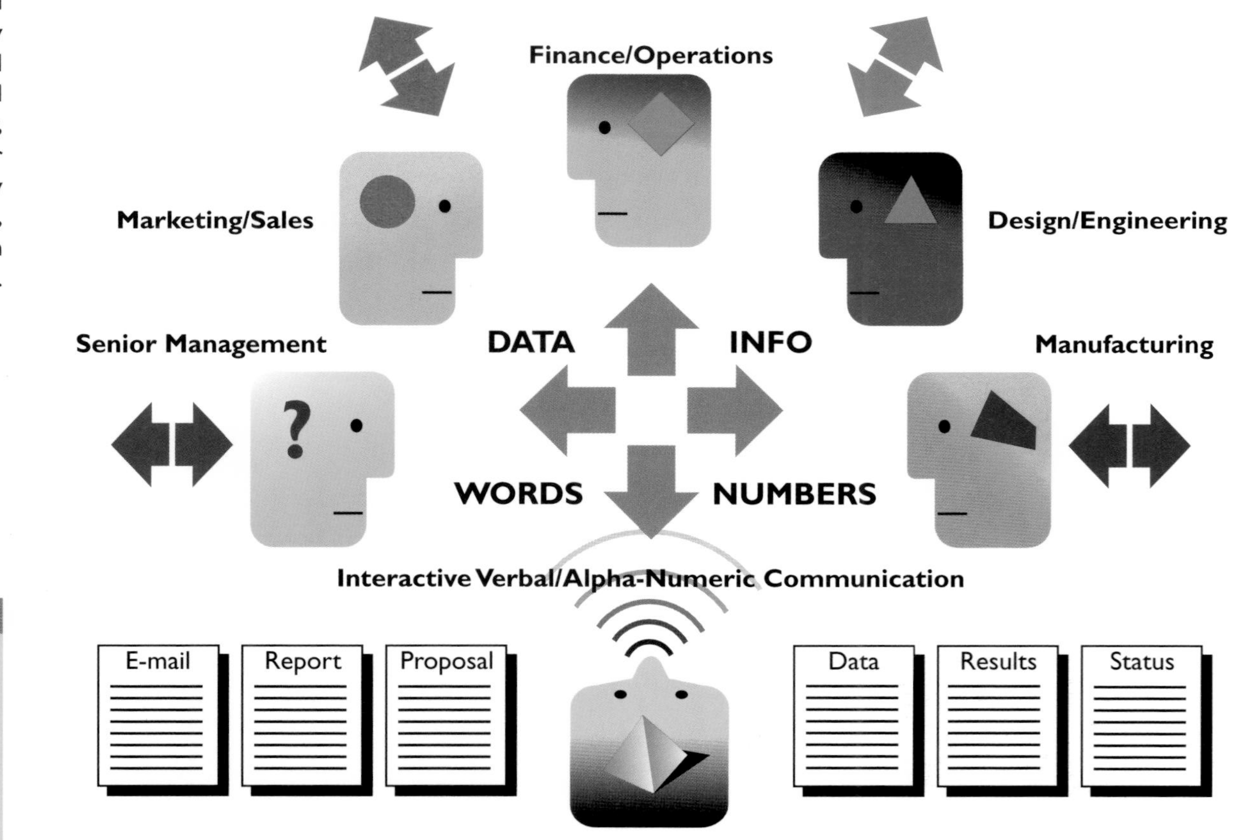

RESULTS & NON-BENEFITS

- > *Confusion and uncertainty*
- > *Inconsistent mental models and images*
- > *Inappropriate actions and reactions*
- > *Lost or wasted time, cost, and effort*
- > *Inappropriate results and outcomes*
- > *Incorrect design intent implemented*
- > *Higher development risks and lowered profitability*

004

NOTES:

Verbal Communication Risk: The downside when a great idea is only communicated verbally

There are dangers in verbally expressing concepts about visual or physical entities. When Jory Olson first proposed his brilliant idea of a round, puck shape for the revolutionary InFocus CableWizard, a data/video projector interconnect product, it was immediately rejected. Why? Because Jory initially said to the development team: "Why not make it round like a hockey puck?" The resultant perceptual images turned off his listeners, and they defaulted to the more obvious rectangular alternatives. It was not until FUSE, a local industrial design firm, was brought in to execute 2D and 3D sketches of circular versus rectangular product shapes that Jory's audience recognized the value of his idea. The original concept, fully conceived in Jory's mind, was right on. But the images conjured by his audience from the first verbal description nearly killed it. Simply visualizing the "puck" concept, with details of form and features, made the communication (and success) difference.

TOOLS, TACTICS & TALENT

> *Visualize, simulate, sensorize, realize*
> *Utilize skilled visualizers as support*
> *Execute enough detail for clarity*
> *Use appropriate level of visual sophistication*

LESSONS TO LEARN

> *Verbal-only metaphors can fail to communicate proper mental models*
> *Communicators' ideas must be transmitted to others clearly*
> *Verbal-only communication can be disastrous to good ideas*
> *Visual/physical ideas should be communicated in a visual/physical manner*

Jory A. Olson: Principal Engineer, InFocus Corporation, Wilsonville, Oregon

Jory Olson received his B.S. in Electrical Engineering and has since worked on numerous projects, including satellite modems for Motorola, whose satellite was the first to have a demodular on-board. Jory can often be found interacting with his mechanical engineering buddies, which helps him find elegant solutions to difficult human interface problems. Jory is very knowledgeable about audio design and musical instruments, and he consults with the Taubman Institute of Piano. In his spare time, Jory can be found restoring his 1925 Colonial Revival home, hiking in the Columbia River Gorge, reading, or playing the piano.

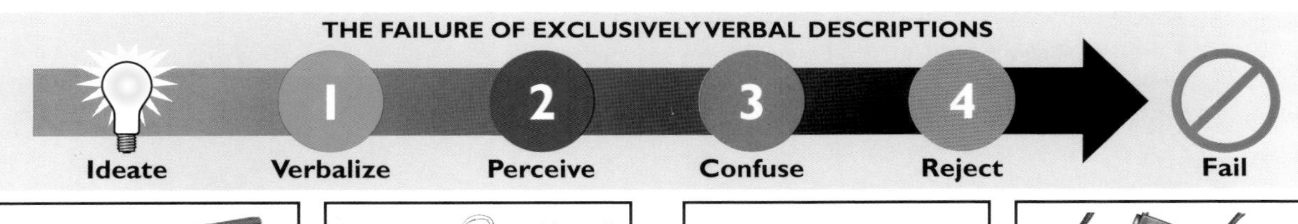

Hockey puck verbal description

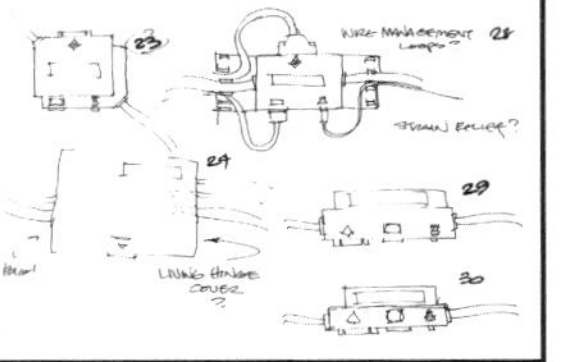
Rectangular version options

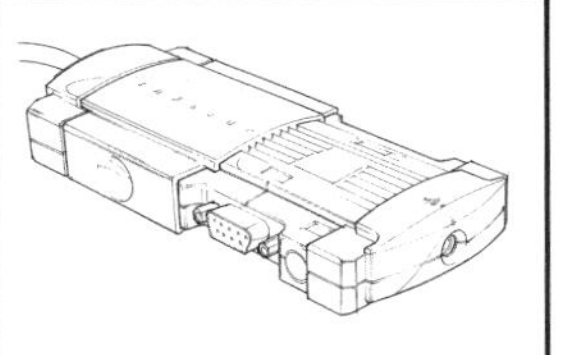
Rectangular design sketch

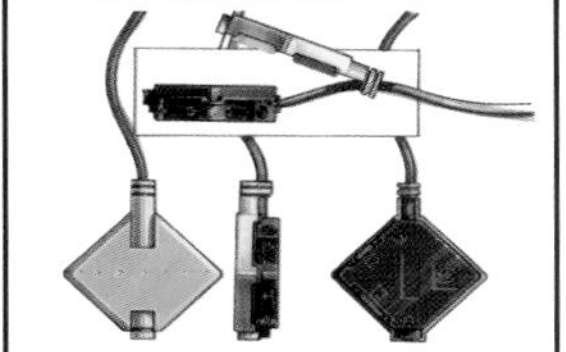
Further rectangular designs

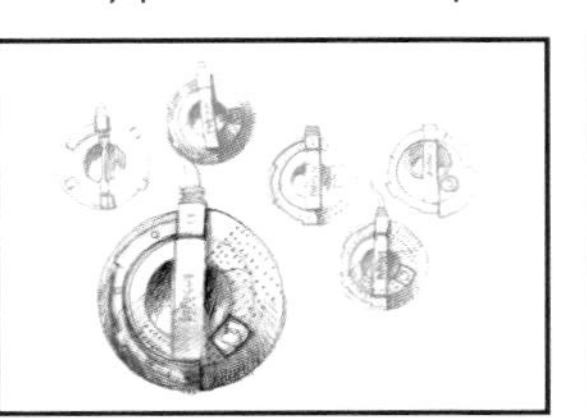
Early round ideation sketches

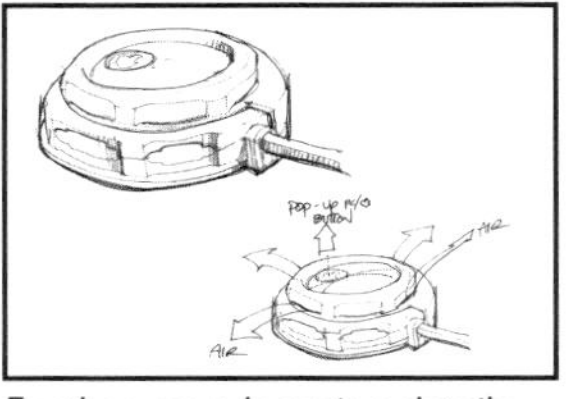
Further round version details

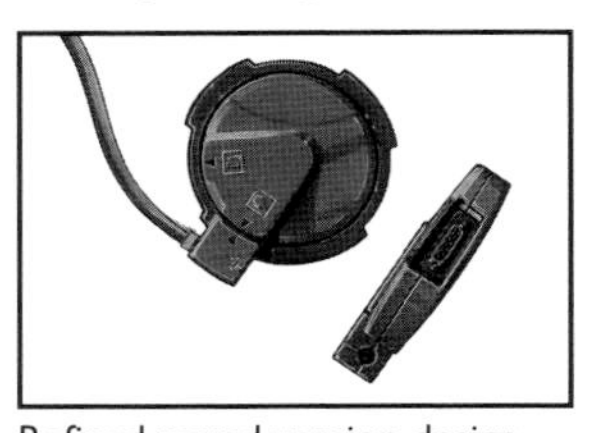
Refined round version design

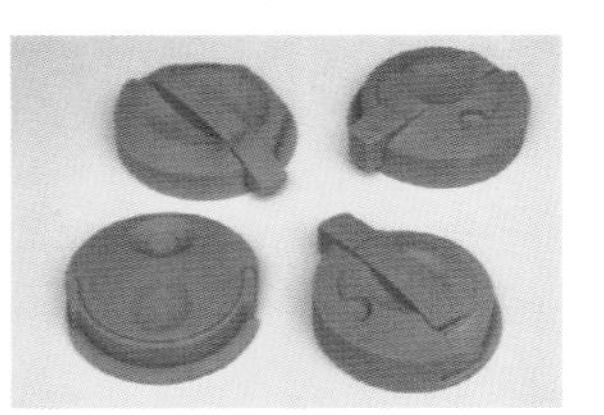
Round foam mockup designs

Successful "puck" design in production

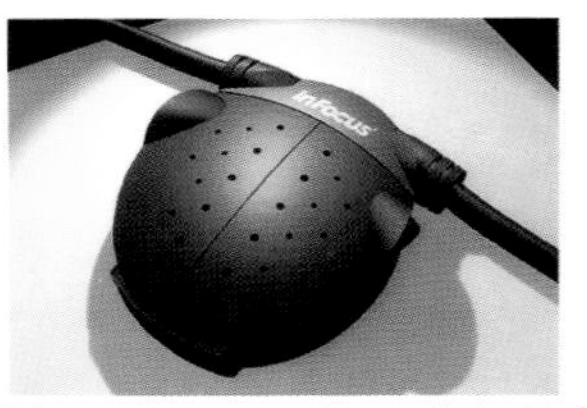

Next generation design ("The Tick")

NOTES:

"Without deviation, progress is not possible."
Frank Zappa

SENSORIZED™ INTERACTION

Productive idea communication using visualization and simulation

Clear communication uses language that appeals to the appropriate senses and is also a key to innovation. Using multiple media appropriately will optimize understanding and results. Bill Roberts helped coin the term Sensorization™: presenting and communicating an idea in the manner most suited to the sense(s) it is to be evaluated with. Auditory ideas should be presented to the ears for hearing, not only by graphs or data. Ideas requiring visual evaluation should be presented visually. The same goes for kinesthetic, tactile, or olfactory ideas. Verbal or alpha-numerical data is appropriate and useful to a certain extent, but one can't "hear" a graph of the frequency response of an audio speaker. Sensorized™ communication and interaction with multiple modes of simulation have a better chance of creating perceptual models that are uniform and clear. This will lead to better actions, planning, decisions, results, and productivity, especially in group environments.

William W. Roberts: Founder/Principal, VisuaLogos, Portland, Oregon

Bill is founder and principal of VisuaLogos, a company providing visual communication services to a broad range of clients. He holds a masters degree in Communication from Pepperdine University. Bill has served as a partner in a financial services marketing consultancy, helping clients like Citibank, Bank of America, and Fidelity Funds communicate effectively with their investment customers. He is a veteran graphic designer and copywriter with a strong background in advertising.

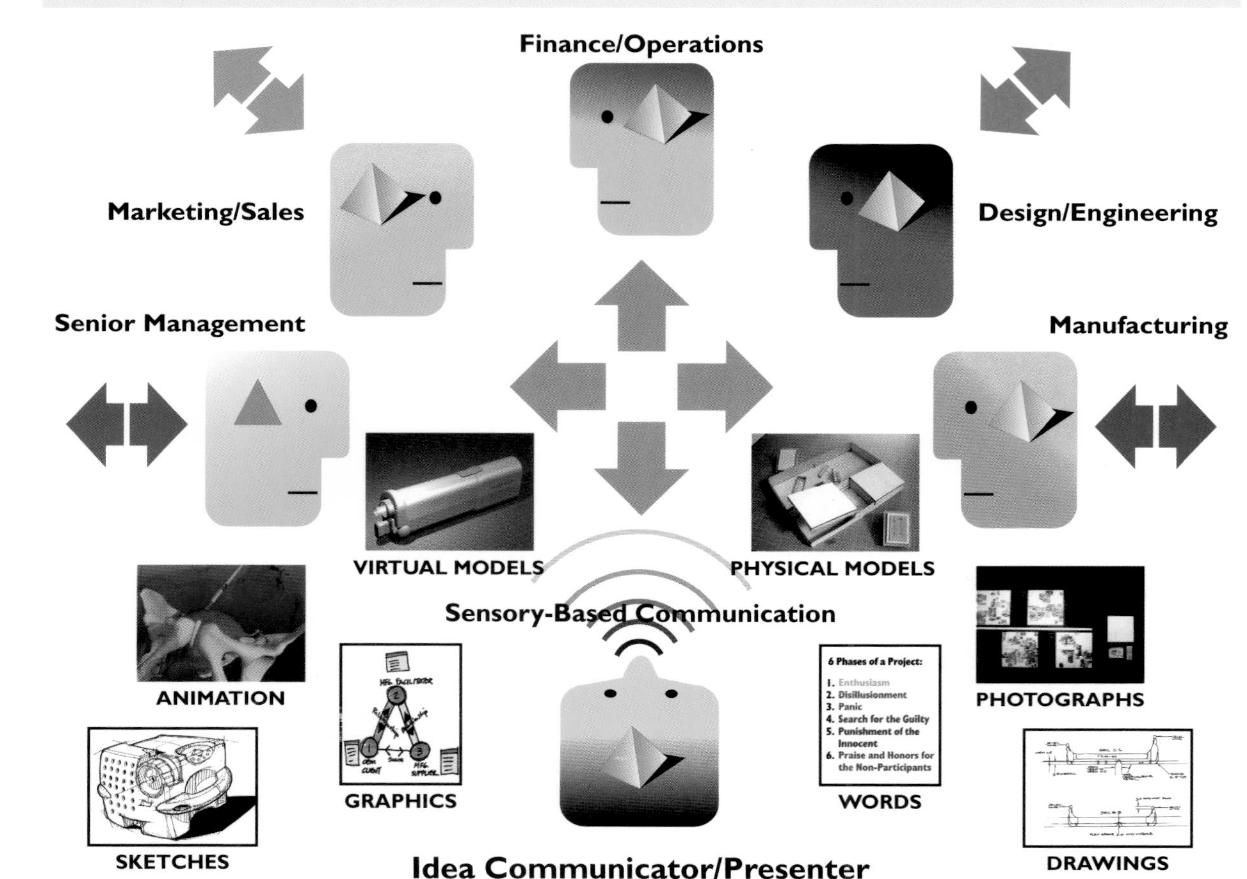

TOOLS, TACTICS & TALENT

> *Use of appropriate 2D and 3D simulations*
> *Talents of 2D and 3D visualizers/simulators*
> *Presentation to appropriate audience senses*
> *Sensory interaction of audience to simulations*
> *Appropriate capture of audience response*

RESULTS & BENEFITS

> *Improved communications and interactions*
> *Optimum actions/reactions to ideas*
> *Better resolution of issues presented*
> *Productive and efficient use of time*
> *Increased idea value and profitability*

NOTES:

Physical Simulation Interaction: Sensorized communication leads to collaborative innovation results

The product marketing group for a high-technology company proposed an innovative product interconnection system that seemed plausible on paper. However, it seemed to some that it may have potential human interaction problems. To test viability early, Mark Schoening and his team at FUSE were brought in to mock up the system and facilitate interaction. The mockups were made of simple materials and constructed to simulate the system components and interconnection. During a group interaction session, it was found that the proposed system could be interconnected incorrectly, and the complex mix of connectors and cables was confusing to the user. By interacting and exploring with the physical mockups, at least two viable alternative designs were discovered. In addition, an engineer passing by was immediately engaged by the mockups and saw another alternative solution that no one else had seen.

TOOLS, TACTICS & TALENT

> *Simple schematic system simulation*
> *Inexpensive 3D mockup materials and processes*
> *Group sessions with hands-on interaction*
> *Involvement of cross-functional teams*
> *Appropriate visual documentation of results*
> *Openness and visibility of process to others*

RESULTS & BENEFITS

> *Quick identification/elimination of poor solutions*
> *Discovery of fresh ideas and innovations*
> *Engagement of others outside of base group*
> *Early design optimization via collaboration*
> *Clarity of system function, use and application*
> *Uniformity of system understanding*

FUSE, Inc.: 1415 SE 8th Avenue, Portland, Oregon 97214

FUSE is an industrial design and development consultancy that engages clients and projects in consumer products, medical devices, transportation, and computer/information equipment. FUSE's strength is quickly and creatively conceptualizing innovative product solutions and turning them into real products. Its list of clients includes well-known organizations such as Anthro Corporation, Herman Miller, InFocus, Intel, Nike, and S3/Diamond. FUSE has won numerous design awards and recognition for its ability to advance products from concept to reality.

Simple wood/foam component mockups

Collaborative Innovation Process

KINESTHETIC SENSORIZATION RESULTS

1 2 3 4 $

Ideate | Simulate | Engage | Collaborate | Design | Profit

Initial system complexity problem discovery

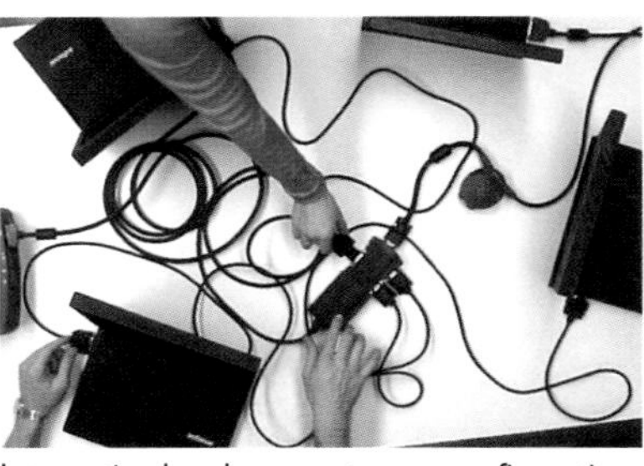
Interactive hands-on system reconfiguration

Ideating several viable alternative solutions

Easy engagement of outside collaborators

NOTES:

INNOVATION CHAMPION

> "Any darn fool can make something complex; it takes a genius to make something simple."
>
> Woody Guthrie

Fostering innovation and creativity with a visionary leader

Crucial to facilitating and maintaining innovation in a development environment is having a resident champion. Such a person takes up the torch for the tools, tactics, and talent necessary for success of the creative process. Champions must be passionate about promoting and encouraging out-of-the-box thinking. Often viewed askance by the conservative business contingent, the innovation champion must be willing to take risks and stand up for whom and what they represent. Innovation methodologies often appear silly and counterproductive to the average corporate traditionalist. Those who must execute creative work require a visionary leader to promote their activities if they are to remain focused and productive. The successful innovation champion must be well-versed in the ways and means of the creative process and be able to foster patience and acceptance within senior management.

TOOLS, TACTICS & TALENT

- > *Creative environments and work spaces*
- > *Ideation and brainstorming processes*
- > *Off-the-wall thinking methodologies*
- > *Value visibility of activities*
- > *ROI communication to management*
- > *Encourage, mentor, and support team*

RESULTS & BENEFITS

- > *Focused creative productivity*
- > *Encouraged and motivated staff*
- > *Visibility of innovation processes and results*
- > *Improved funding of creative processes*
- > *Better, more innovative product solutions*
- > *Improved profitability and bottom line results*

Mark Sylvester: Co-Founder and Ambassador, Alias|Wavefront Corporation, Santa Barbara, California

Mark is one of the founders of Wavefront Technologies and one of the first animation software developers. He initially helped to develop the Advanced Visualizer, a 3D computer animation system first used at Universal Pictures. Mark is one of the co-designers of Composer, a digital compositing system, currently used in most major motion pictures produced in Hollywood. He now serves as Ambassador for Alias|Wavefront, where he works as a liaison between customers and the company to ensure a close relationship between artists and developers. Prior to founding Wavefront Technologies in 1984, Mark was a professional chef and helped build several restaurants in the Santa Barbara, California area.

CREATIVE INNOVATION TEAM

INNOVATION CHAMPION

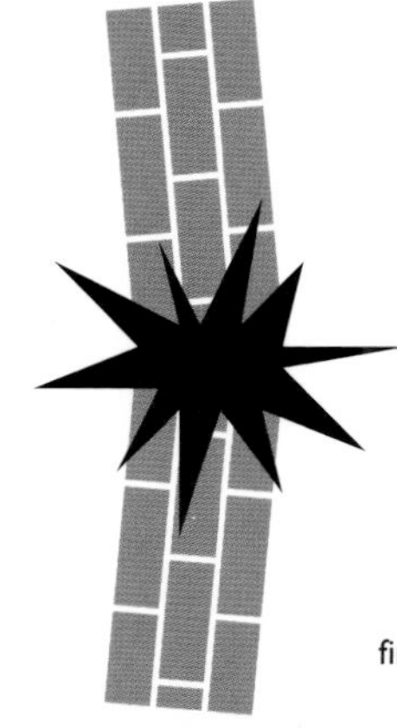

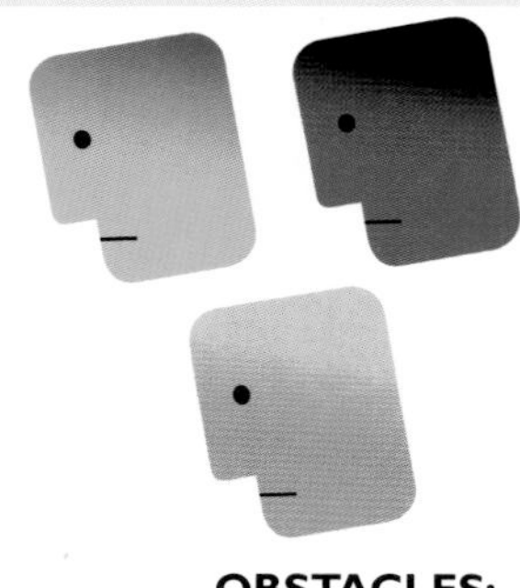

OBSTACLES:
Poor management or leadership, lack of vision, physical constraints, financial limitations, resistance to change, ignorance of creative process, etc.

MARK'S OBSERVABLE BEHAVIORS OF INNOVATION LEADERS

1. Demonstrate care and commitment to clients in our words and actions.
2. Be an enthusiastic student of our industry, our technology, and our current and future clients.
3. Practice teamwork—cooperate with, care for, support, and respect each other.
4. Use open, honest, constructive communications. Don't leave things unsaid.
5. Listen, clarify, and empathize.
6. Keep commitments.
7. Seek and support alignment.
8. Focus on high-value, high-leverage activities.
9. Be accountable for the results we produce or fail to produce.
10. Work hard, laugh in the halls, and get great stuff done.
11. Act like you own the company.
12. Celebrate success!

NOTES:

Ease of Use Champion: Passionate advocacy for simplicity of use

One of Bill Buxton's favorite sayings is that "Making things hard is easy. Making things easy is hard." This is the great challenge of the user interface designer: making a product as easy, friendly, and compelling to use as possible. User studies have shown many times that people gravitate toward products that are easy and enjoyable to use, and will choose them over products that are difficult to use but that may offer more functionality. One reason for this is that regardless of how many features or functions a product may have, if they are not easily accessible to the customer, they become essentially worthless. They are, in fact, frustrating for the user who has paid for this functionality but cannot capitalize on it. As a champion of the user at both SGI and Alias|Wavefront, Bill Buxton is passionate about developing new technologies that enhance functionality for his customers and about making those features as easy to use as possible.

William Buxton: Chief Scientist, Alias|Wavefront and Silicon Graphics, Inc. (SGI), Toronto, Canada

Bill is a designer and researcher who specializes in the human aspects of technology. He is chief scientist of Alias|Wavefront, Inc. and its parent company, SGI, as well as an associate professor in computer science at the University of Toronto. After receiving a Bachelor of Music degree, Bill completed an M.Sc. in computer science on computer music at the University of Toronto. He joined Alias|Wavefront in June, 1994. In addition to being an avid equestrian, Bill is co-owner of a contemporary art gallery in Toronto.

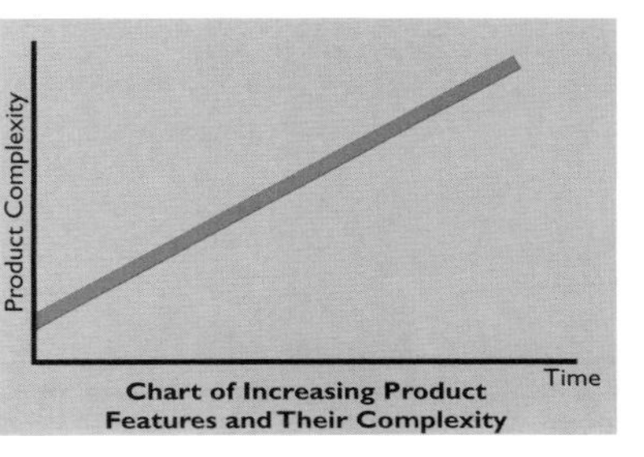

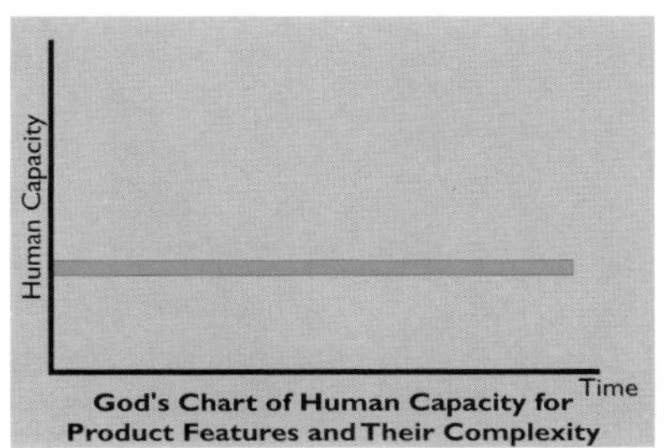

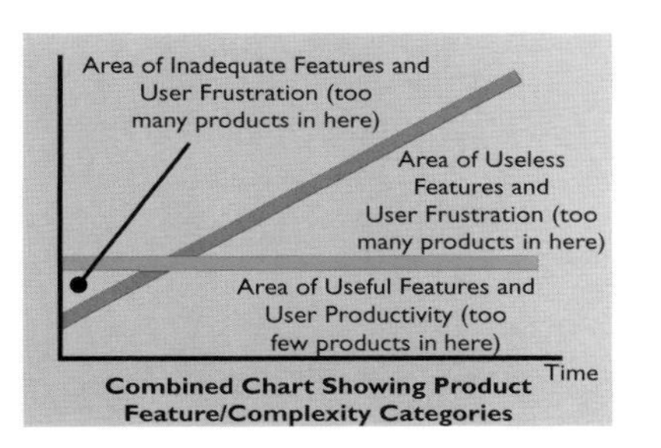

Buxton's User Reality Charts

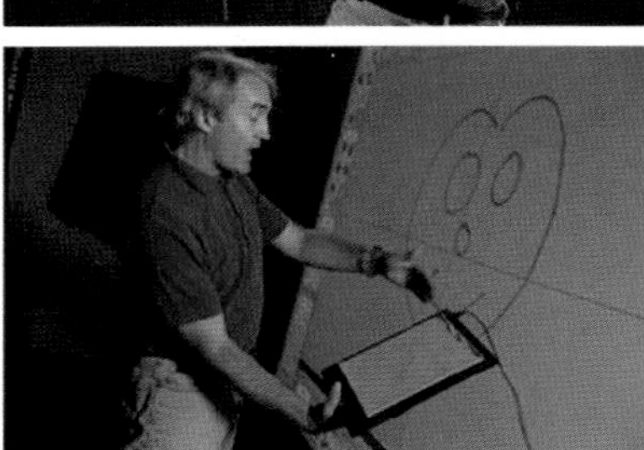

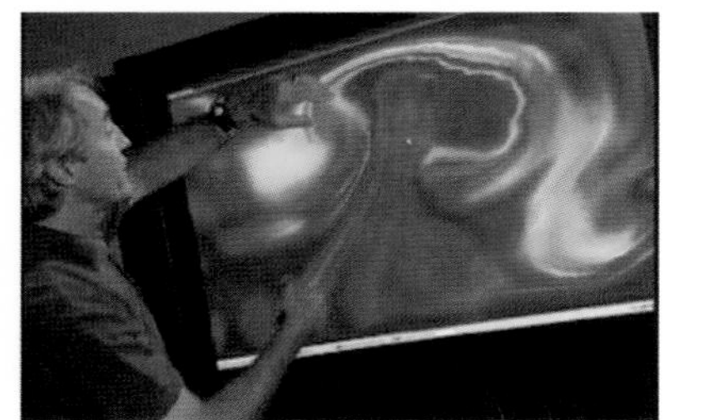

Examples of Bill's many activities in new product ideas, cutting-edge technologies, interaction design, product usability, and human factors

NOTES:

INNOVATION WORKPLACES

"One of the advantages of being disorderly is that one is constantly making exciting discoveries."

A. A. Milne

Establishing work environments that encourage creativity

The personal work environment is highly important to optimizing both personal and corporate productivity. An enterprise that wishes to maximize the innovation and work quality of its development personnel would do best to provide work spaces that are conducive to such activity. When workplace ambiance is optimized for creativity and productivity at all levels, everyone benefits. Providing such environments need not be complicated nor expensive. The key is to permit individual design, implementation and maintenance of personal work space as much as possible. The wrong approach is to assume that providing expensive cookie-cut cubicles will satisfy everyone's needs. People need to control the conditions of their own work area. For some, this may be a highly pristine and regimented work space. For others, it may be simple and sparse. For still others, a flexible work space may appear chaotic, but is perfectly appropriate for that user.

TOOLS, TACTICS & TALENT

> *Be flexible to needs of occupant*
> *Personal ambiance within social acceptance*
> *A place of comfort and friendliness to the user*
> *Accommodates both interaction and privacy*
> *Open to adequate level of personal expression*

RESULTS & BENEFITS

> *Increased performance and productivity*
> *Increased job satisfaction*
> *Maximum time spent at work*
> *Increased innovation and creativity*
> *Inspiration and experience for others*

Creative Environments Foster Creative Minds

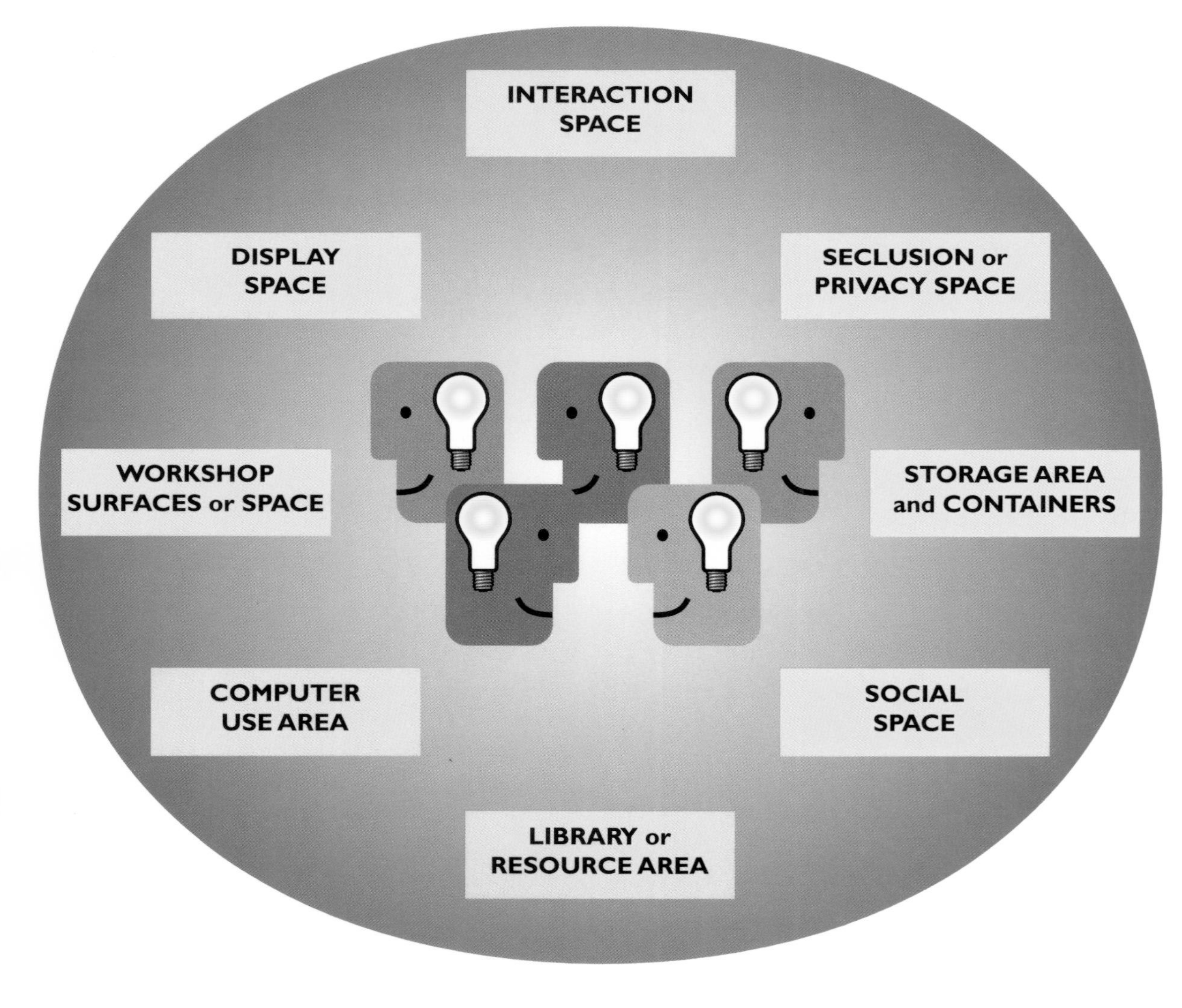

NOTES:

Creative Work Areas: Unique workspaces that encourage innovation

With headquarters located in several buildings and office spaces along High Street in Palo Alto, California—the heart of Silicon Valley—IDEO's facilities are a true delight. Its studios, work spaces, and offices epitomize environments promoting innovation and creativity. IDEO's model shops, design labs, work areas, conference rooms, and interaction spaces are all unique and interesting. Artifacts of design abound everywhere: hanging from the ceiling, tacked to the walls, laying around work spaces, and tucked into nooks and crannies. Though it is clear that each work space is highly individualized, the comprehensive feel is one of collective collaboration and interaction. IDEO has clearly recognized the value of the work environments designed for productive innovation.

TOOLS, TACTICS & TALENT

- > *Flexibility of space and materials*
- > *Freedom to individualize*
- > *Tolerance of organized chaos*
- > *Mobile, interchangeable components*
- > *Designed for both interaction and privacy*
- > *Opportunities for relaxation and rest*

RESULTS & BENEFITS

- > *Improved staff performance*
- > *Increased job satisfaction*
- > *Innovative and creative productivity*
- > *Optimized design development*
- > *Attractive recruiting feature*
- > *Showcase for clients/customers/investors*

IDEO: 700 High Street, Palo Alto, California 94301 (and worldwide)

IDEO is one of the largest product design and development firms in the world. It was founded by CEO David Kelley in 1978 and has become a major force in product development. IDEO offers a comprehensive range of services from industrial design, product engineering, and concept innovation to full-scale turn-key product development and manufacturing. With its strong Stanford University heritage, IDEO's strength lies in applying the creative process across the entire enterprise. A broad cross-section of professional disciplines, plus multiple locations, studios, shops, and engineering laboratories, enables IDEO to tackle virtually any type of development project.

Four examples of IDEO design studio and engineering lab workspace environments

Creative IDEO model and prototype workshop environment

A big space for an IDEO Christmas party

NOTES:

War Rooms: Borrowing a military tactic to improve development innovation and productivity

War rooms have been used by the military for centuries to work out battle strategies. In the same way, project "war rooms" are used by product development teams to conceptualize and plan. At RKS Design, war rooms were important enough to consciously incorporate into their facility in southern California. Three such rooms were included with the essential features for proper utilization and security. These rooms become the "homes" for major projects throughout their life at RKS. Much of the project material remains in these rooms for ongoing individual, client, or team review. Included are components that support immersion in ideation, visualization, interaction, and communication.

TOOLS, TACTICS & TALENT

> *Designated, secure room for each project*
> *Ideation, interaction, and presentation features*
> *Computer, communication, and Internet tools*
> *Areas, resources, and materials for design*
> *Amenities conducive to collaborative innovation*
> *Essential components:*
- *white boards (preferably electronic)*
- *bulletin/tack boards and display areas*
- *projection systems (digital, video, and slide)*
- *work tables, tools, and furniture*
- *project materials and visuals storage*

RESULTS & BENEFITS

> *Home for major project activities and materials*
> *Intellectual property protection and security*
> *Efficient and distraction-free productivity*
> *Focused management interaction*
> *Team-building environment*
> *Improved overall project performance*

RKS Design, Inc.: 350 Conejo Ridge Avenue, Thousand Oaks, California 91361-4928

RKS Design, Inc.: 350 Conejo Ridge Avenue, Thousand Oaks, California 91361-4928

Founded by president and CEO Ravi Sawhney, RKS Design is an industrial design and product development consultancy located in a custom-designed facility in Thousand Oaks, California. The firm employs a variety of professionals that includes designers, engineers, model makers, and administrative and support staff. RKS Design engages in a broad range of projects, from toy and entertainment design to high-technology products such as computers and audio equipment. They have received numerous international awards and recognition for their design and development work. RKS clients include Hewlett-Packard, Apple Computer, Panavision, Canon, and Amana.

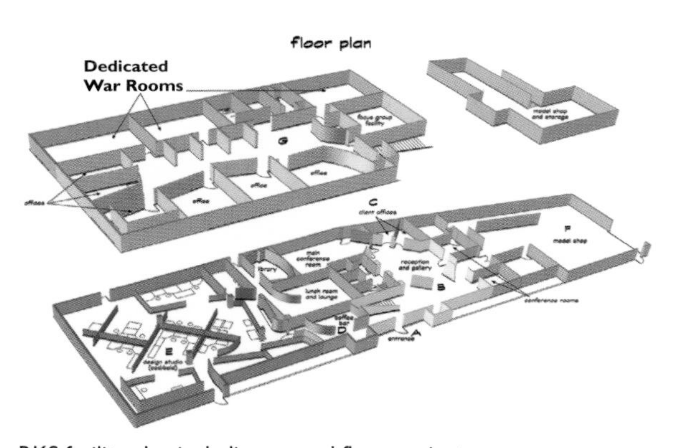

RKS facility plan including second floor project war rooms

Project development team interaction session

Development team war room work session

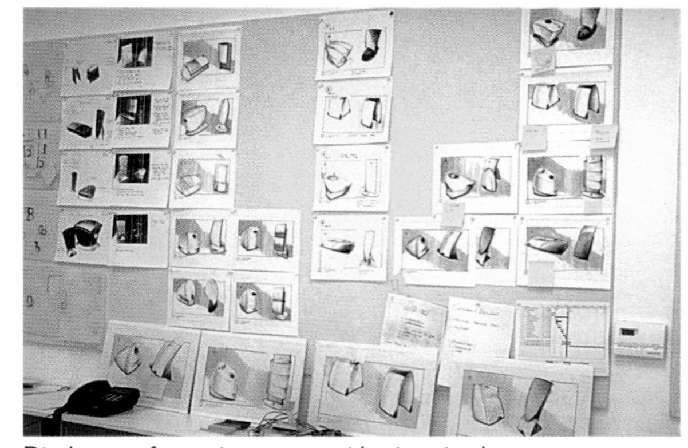

Display area for project concept ideation visuals

NOTES:

Development Ambiance: Innovative workplaces for clients and staff

Creating a positive environment for customers and staff can be very rewarding, both emotionally and economically. RKS Design has separated its product development facility into two major sections, creating a friendly working and interaction ambiance for two important groups: designers and clients. Each group has its own environment in which to interact and work without interference. Development personnel are thus able to work as a focused team. RKS Design clients are also provided with their own areas of receptive comfort complete with furniture, interaction space, and refreshments. Also provided are two or three private offices equipped with telephones, computers, and Internet service for client use. Overall, provision is made for both clients and staff to get their work done either independently or interactively.

Ravi K. Sawhney: President and CEO, RKS Design, Inc., Thousand Oaks, California

Prior to founding RKS Design in 1980, Ravi was a professor of industrial design at California State University at Northridge. He is a sought-after speaker on how the power of design can communicate consumer benefits for strategic business advantage. Ravi created Psycho-Aesthetics™, a methodology that discovers the often subliminal visual, tactile, and auditory cues that communicate product values. This RKS proprietary process has led to numerous awards and market successes for RKS clients. Besides running the RKS daily operations, Ravi is also co-founder and board member of Intrigo, a mobile computing workspace solutions company.

Client reception/display area

Client interaction center

Exclusive client privacy rooms

Secure/restricted design studio section

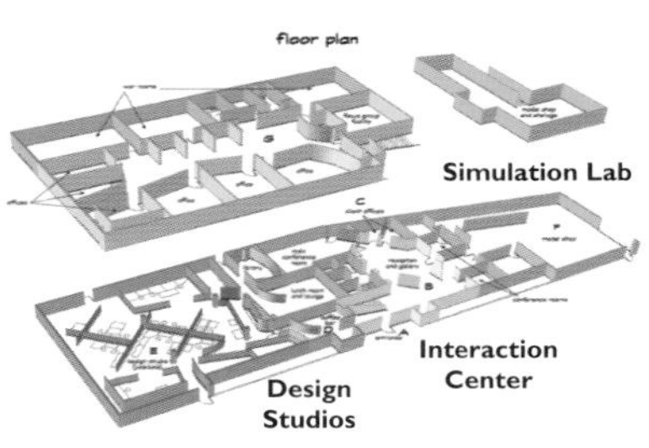

Architect's plan with first floor sections

Design studios and work spaces section

Design staff amenities

Flat organizational structure

TOOLS, TACTICS & TALENT

- *Definitive focused spaces for staff designers*
- *Separate friendly reception spaces for clients*
- *Appropriate refreshment and relaxation amenities*
- *Appropriate work spaces for staff and clients*

RESULTS & BENEFITS

- *Improved productivity of all*
- *Minimized staff and client distractions*
- *Welcomed, comfortable, and productive clients*
- *Positive staff/client relationships*
- *Optimum client experience provider*

NOTES:

> "A management obsessed with productivity usually has little patience for the quiet time essential to profound creativity."
> Gordon MacKenzie

VALUE OF PEOPLE

The organization that values and supports its people will achieve optimal success in innovation

"There is no mystery to creating innovative design. It's all about respecting an individual's unique capabilities and supporting their universal desire to contribute. If you get great people, they will attract great projects, which is great fun. Designworks/USA was started simply by the passionate love of the design process with a total respect for the needs of the individuals. The basic unification of thought was that our employee mix had to be multicultural and multi-talented, our client base multinational, and our services multidisciplinary. This combination provided a cross-fertilization that fostered continuous learning, stimulation, and inspiration. The office was able to seamlessly apply creative insight from one designer to another, from team to team, from project to project, which accelerated as the knowledge and breadth increased. This philosophy indelibly instilled an inner sense of belief in each person that transferred to our clients in long-lasting relationships. These clients were really person to person relationships that carried the company through changing climates."—Chuck Pelly

TOOLS, TACTICS & TALENT

- > *Recruit internationally and cross-functionally*
- > *Diversify projects and teams*
- > *Establish international design sites*
- > *Create an environment of security and acceptance*
- > *Accept failure as a part of success*

RESULTS & BENEFITS

- > *Advances technological processes and resources*
- > *Stabilizes staff and sustains creativity*
- > *Optimizes global marketing strategies/partners*
- > *Develops internationally marketable products*
- > *Maintains client confidence and relationships*

Charles W. Pelly: Founder/Consultant, Designworks/USA (BMW), Newbury Park, California

Chuck Pelly, founder of Designworks/USA, brought this subsidiary of BMW from an operation in his garage to being one of the premier design establishments in the world. He fostered creative thinking, emphasized people, and utilized several aspects of design to create the unique environment at Designworks/USA. Chuck currently has his own independent design and creative consulting group, Pelly Design Management, and is an active mentor and lecturer at his alma mater, Art Center College of Design in Pasadena, California. He received the highest achievement award of Fellow from the Industrial Designers Society of America and has held many offices in that organization, including president.

Anniversary marble and aluminum plaque embedded in the foyer floor at Designworks/USA by Chuck Pelly

NOTES:

Multicultural Teams: Development teams for global markets require international input

In today's highly competitive markets, the ability to sell products within a multitude of countries and cultures is critical. Marketing requires products to appeal and be adapted to multiple users from diverse cultures and backgrounds. Designing products that do this is optimized by familiarizing people on the product development teams with those cultures, as is customary at Designworks/USA. By assembling teams that are multinational, multicultural, and multiethnic, the challenge of globalized design is reduced. Research shows that aspects of a product's features such as color, form, packaging, etc., can be highly positive or devastatingly negative to its success in a particular culture or nation. Purchase influence significantly depends upon the cultural nuances of the country of sale.

Designworks/USA (BMW): 2201 Corporate Center Drive, Newbury Park, California 91320

Designworks/USA, a subsidiary of BMW, is an internationally recognized design consultancy that specializes in automotive, transportation, and product development. Its headquarters in southern California's "technology corridor" maintains extensive modeling shops, engineering labs, and industrial design studios, with a staff of designers, engineers, model makers, and support professionals. Some of Designworks/USA's most prominent clients include Allergan, Nokia, and Hewlett-Packard, plus its parent, BMW, for which it does extensive automotive design.

NOTES:

"There is nothing so wasteful as doing with great efficiency that which doesn't have to be done at all."

Anonymous

PARA-MANAGEMENT MODEL

A management support model that frees managers to do what they ought to be doing

When technical contributors show leadership and vision in their work, they are often promoted to management. Unfortunately, they may move to a position with responsibilities that are contrary to what they do best. Their creativity and leadership are soon displaced by management duties. Revenue-generating value is lost to bureaucracy. Typical technical track options offered as a solution don't really work, since real power is still with management. So how can one accomplish the bureaucratic/administrative work while facilitating visionary product development and directly participating in revenue-producing leadership? An option: the para-manager model. This paradigm utilizes the support of a dedicated, capable para-manager to execute administrative management duties so that the visionary innovation leader can be focused productively.

TOOLS, TACTICS & TALENT

> *Emphasize facilitation and innovation*
> *Use competent, talented para-manager*
> *Leverage skills of para-manager*
> *Build close para-manager relationship*
> *Delegate administrative tasks*

RESULTS & BENEFITS

> *Manager motivation and retention*
> *Increased innovation performance*
> *Efficient administrative execution*
> *Optimized departmental performance*
> *Improved bottom line revenue*

Typical Promotion to Management Position vs. Para-Management Model

Explore & Risk
Misc.
Vision & Leadership
Designs & Concepts
Innovation
Ideas
Development

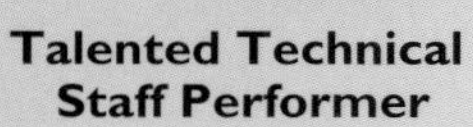

Talented Technical Staff Performer
(Revenue-Generating Development)

THE SOLUTION:

PROMOTION USING PARA-MANAGEMENT MODEL

Creative, innovative technical staff member is recognized for leadership and promoted into true leader/manager role, with para-manager support designed to fully utilize the visionary and creative talents of the leader/manager

THE PROBLEM:

TYPICAL MANAGEMENT PROMOTION MODEL

Creative, innovative technical staff member is recognized for leadership and promoted into a management position that allows little opportunity and time for employment of visionary/leadership talents

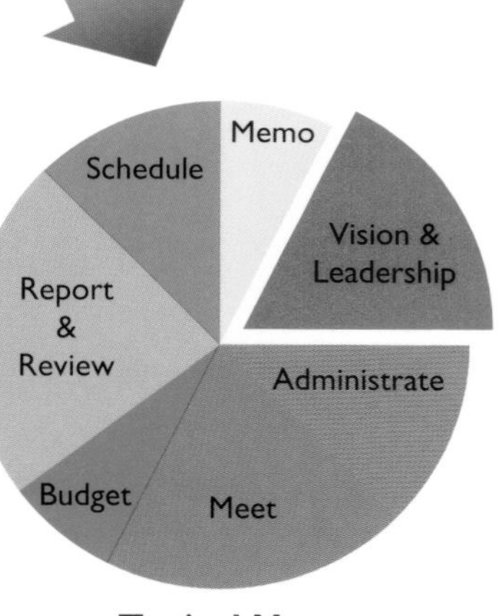

Typical New Management Functionality

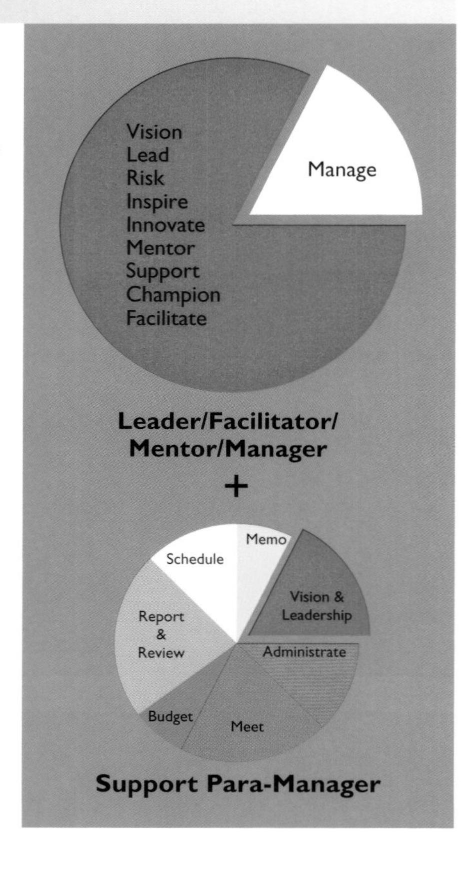

NOTES:

Para-Manager in Action: A management model that works in a variety of applications

Critical to the success of the para-management model is that both the manager and the para-manager should have the right characteristics. Judy Dancer is an example of a natural para-manager. She was a technical trainer, moved to a project coordinator position, and is now an engineering projects manager. Itemized below are some of the most important characteristics and personal traits necessary for making this model work well. The essence of the para-manager role is to focus on creative and innovative execution: accomplish the management objectives that the visionary manager delegates. In Judy's aggressive work environment, this means not simply getting the job done, but making it better, pushing back, and taking initiative as well as risks. She does whatever is necessary to successfully advance the vision of the leader, the department, and the company.

VISIONARY LEADER ATTRIBUTES

> *Dedication to innovation*
> *Ability to delegate and feel great about it*
> *Teacher, trainer, mentor, and supporter*
> *Appreciation for detailed, thorough, quality work*
> *Rewards and praises innovation and productivity*

PARA-MANAGER ATTRIBUTES

> *Satisfied being number two*
> *Willing to do what it takes to execute properly*
> *Takes instruction well and positively*
> *Goes out of way to amplify and innovate*
> *Attention to detail and nuance*
> *Persistence in resolving problems and issues*
> *Executes impeccably in service to others*
> *Builds own relationships and leadership role*

Judy Dancer: Manager, Engineering Projects, InFocus Corporation, Wilsonville, Oregon

Judy Dancer is a key player at InFocus, where she identifies and analyzes engineering responsibilities to minimize duplication and wasted time. Judy believes an engineer's job is to design full time, and this must be supported in a variety of ways. Judy began at InFocus as Project Coordinator II and was previously Special Projects Coordinator for an environmental test equipment manufacturer. Judy was also the founder/owner of a food company in Seattle, Washington. She has an A.S. degree in medical technology and a B.S. in business administration.

Appropriate position of the para-manager in the organizational structure

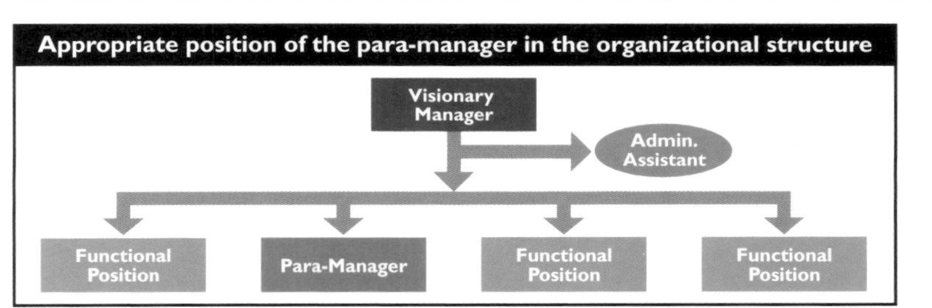

Working process of leader and para-manager

Visionary Leader
Para-Manager
GOAL

Group interaction facilitation

Meeting management and data gathering

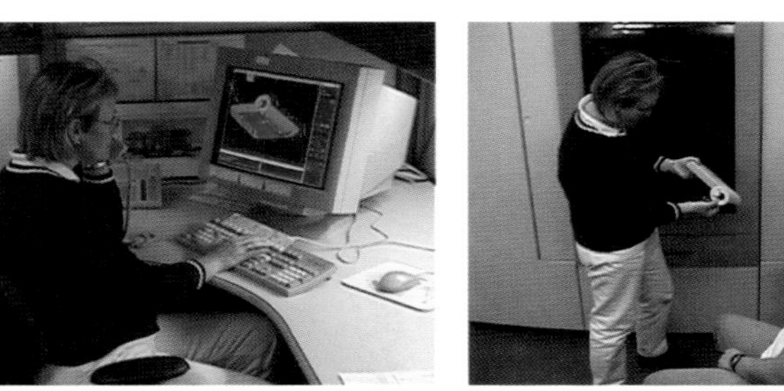
Project research and information analysis

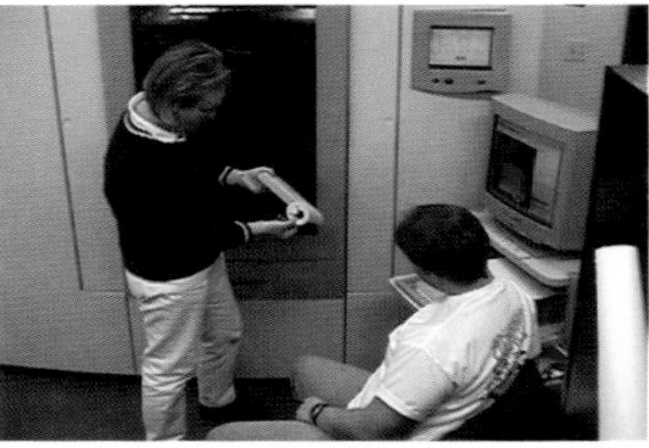
Project coordination and support

Group recreation planning

NOTES:

"This 'telephone' has too many shortcomings to be seriously considered as a means of communication. The device is inherently of no value too us."

Western Union internal memo 1876

The challenges of coordinating multiple global design sites in a fast-paced world market

Design has been key to propelling Nokia to the top as the world's largest mobile phone maker. Personalization and ease of use has been at the heart of its design and marketing philosophy. Because Nokia manufactures and markets phones around the world, it makes sense to build a global design team. Maintaining the highest quality, innovative design solutions for brand products is the role of Nokia vice president and chief designer Frank Nuovo, his senior managers, and his design team. Flexibility is at the heart of the team's success. Coordinating meetings to accommodate all time zones is a challenge, and the Nokia team has to act globally in its approach. For example, creating Nokia's unique graphical mobile phone covers required extensive research and development in the area of decoration technologies, and only a unified global team effort was adequate for the task.

TOOLS, TACTICS & TALENT

> *Single visionary design champion/leader*
> *Multiple creative design studios in key locations*
> *State-of-the-art global interaction technologies*
> *Interactive team approach to development*
> *Frequent visits and mentoring of teams by leaders*
> *Top management committed to world-class design*

RESULTS & BENEFITS

> *World-class product design and innovation*
> *Collaboration of productive global team*
> *Multinational market solutions*
> *Global coordination of manufacturing and design*
> *Market leadership and high profitability*

Nokia, Inc., Nokia Design Center: Calabasas, California

Originally a company that manufactured paper, Nokia has established itself as a world leader in the highly competitive market of digital communication, especially mobile and cellular phones. Nokia originally developed several of the now standard features of modern mobile phones. Nokia is also a powerful force in wireless telecommunications, and it continues to recognize and capitalize on technological opportunities through innovation, simplicity, and style.

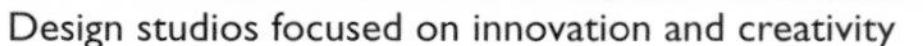

Design studios focused on innovation and creativity

Collaborative team approach to design

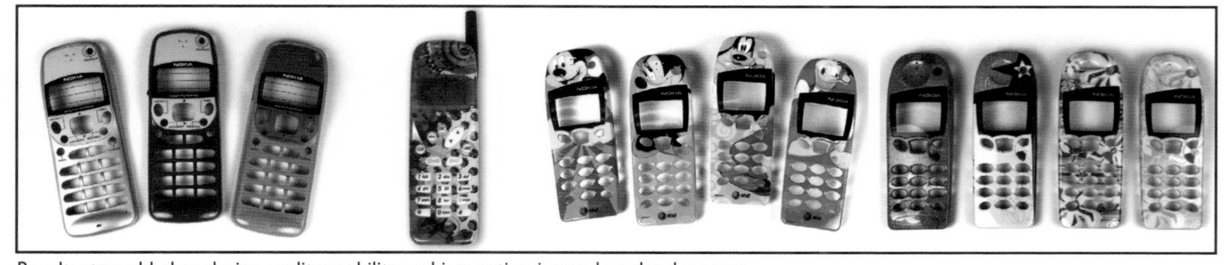

Resultant world-class design quality, usability, and innovation in product development

NOTES:

Personalizing Products: Designing with global input provides world-class results

Nokia worked hard to pioneer personalization with communication products. Whether it be colorful and expressive face plates for the Nokia 5100 and 3200 series or the sleek, sophisticated design of the 8800 series, personalization is at the core of Nokia's design. First came simple changeable colored face plates; then, user-changeable covers. The next step was designing an "engine" which allowed both front and back covers to be removed. Creating a product that was globally available, made in huge volumes, and still very personalizable was a substantial challenge. By using advanced decorating technology, the Nokia 3210 was released in 1999.

Excellence in the creative process requires that alternate solutions be carefully considered. Using hybrid methodologies of hand- and computer-generated artwork and modeling, Nokia's design staff explores many possible solutions. Work is refined as part of a concurrent engineering process which includes review and refinement using 3D CAD. A constant flow of rapid prototype samples helps determine needed functional and aesthetic improvements. The result is a product that changes form dramatically from a traditional to a graphically dominated format within seconds.

Frank Nuovo: Vice President and Chief Designer, Nokia, Inc., Nokia Design Center, Calabasas, California

Frank Nuovo is vice president and chief designer for Nokia and is responsible for all corporate design innovation. Before Nokia, Frank contributed to the design and development work at Designworks/USA, a subsidiary of BMW, and was design director for all Nokia programs while there. Following several years of independent study in art and music, Frank earned his B.S. in industrial design from Art Center College of Design. In addition to being a board member and guest instructor at ACCD, Frank is also on the Business Board of the Finnish National Museum of Modern Art and is a member of the Industrial Designers Society of America. Frank has lectured at several prominent universities and has also been featured in numerous international publications.

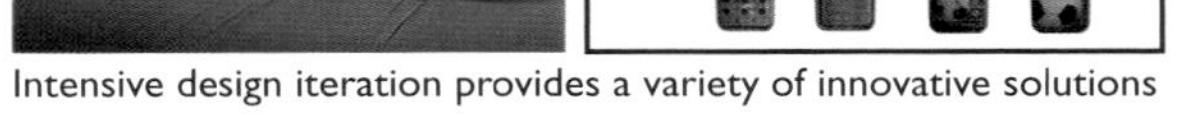

Intensive design iteration provides a variety of innovative solutions

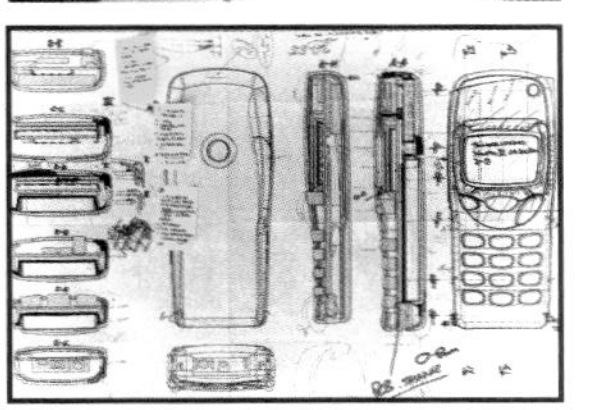

Multiple creative processes result in thorough and quality design

NOTES:

MANAGING OUTSIDE RESOURCES

Successfully managing outside resources is crucial to optimum performance

Corporate management often avoids utilizing outside design and development services due to fear of negative cost, productivity, and ROI outcomes. Instead, they choose to drive their internal resources even harder and faster (to the detriment of both projects and people) to avoid using outside resources. Actually, using outside development resources, if chosen intelligently and managed properly, can significantly enhance project performance, product innovation, and corporate bottom line. Many areas of product development such as industrial design, market branding, manufacturing and assembly, and even technology development can be out-sourced in partnership programs that better benefit the corporation and the product end-users. However, an important key factor is the management of those outside resources. Unless in-house management of outside services is dedicated, knowledgeable, and intelligent, failure will result.

TOOLS, TACTICS & TALENT

> *Proven track record of innovative solutions*
> *Adequate staff to perform duties*
> *Innovative/creative mentality and philosophy*
> *Environment that expresses character*
> *In-house management with out-source experience*
> *Support and motivate outside resources*
> *Collaborative team approach with out-sources*

RESULTS & BENEFITS

> *Expanded expertise, resource, and talent base*
> *Improved creativity and innovation factor*
> *Fresh ideas, processes, and talent*
> *Expanded networks and market partnerships*
> *Faster project schedules and time-to-market*
> *Optimized overall development process*

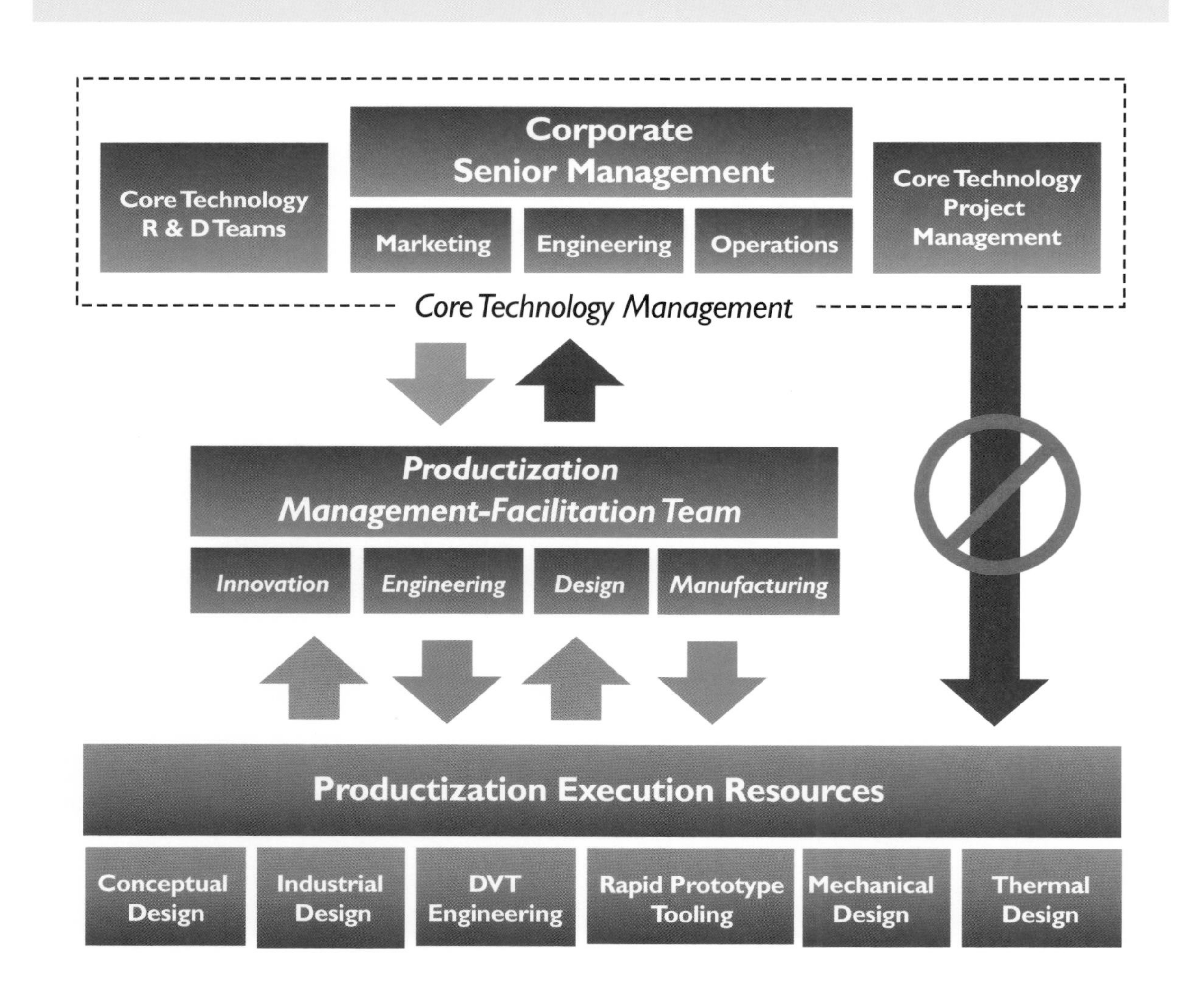

NOTES:

Development Relationships: Outside resource relationships that can be lasting and profitable

When it's time to roll out a new product, companies must decide whether to make do with in-house resources or hire an outside design firm. When the product is mission-critical, though, an outside firm can be a necessity. So it was with Iomega in 1993. In difficult straits with a low stock price, their next product not only had to be successful, it had to save the company. Iomega retained Fitch, a well-respected comprehensive design firm based in Columbus, Ohio. The partnership resulted in the Zip drive, a wildly successful product that not only saved the company but put its primary competitor out of business. Fitch didn't just design a new product, they helped Iomega redesign itself as a company. Successful manufacturer/designer partnerships often result in new corporate identities, marketing approaches, and other larger-than-one-product outcomes. That's why selecting and managing an outside resource firm is so critically important.

Spencer Murrell: Executive Vice President and Managing Director, Fitch, Inc., Columbus, Ohio

Spencer leads product development for Fitch in the U.S. and directs projects for a wide variety of industries, with particular expertise in high-technology consumer products. Specializing in integrating research, design, and engineering, Spencer has directed programs for Kodak, Escalade, Qualcomm, Mitsubishi, Iomega, Checkpoint, Cybex, and Whirlpool. His work has been acknowledged for its excellence by IDSA and has appeared in Business Week, Esquire, Innovation, *and several design books; his name is on numerous international design patents. Spencer graduated from the University of Cincinnati with a B.S. in industrial design and he is a member of IDSA.*

Basic core product development

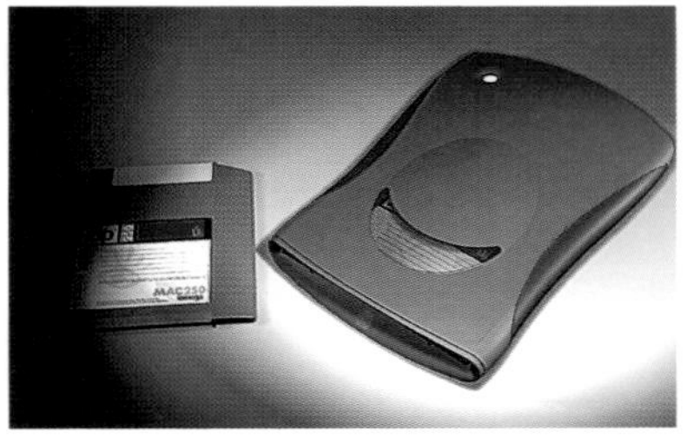

Evolutionary design and development

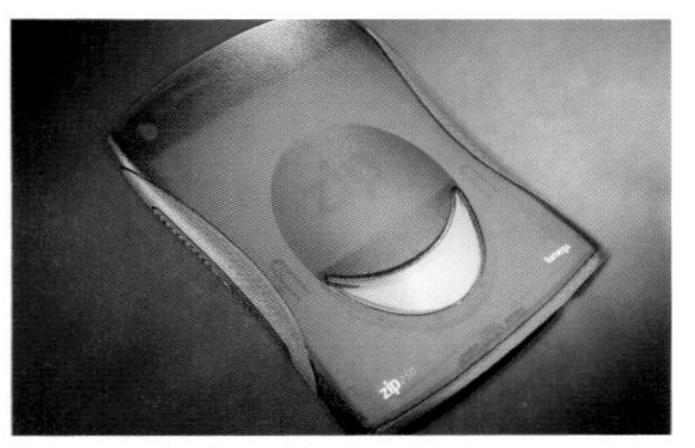

Design of look and feel

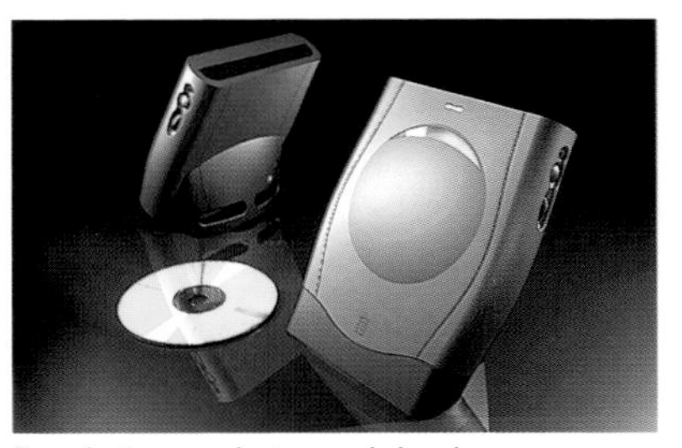

Revolutionary design and development

Usability feature design

Product family design

Design for customization

Packaging design

Exhibit design

NOTES:

> "A common corporate personality disorder: the organization officially lauds the generation of new ideas while covertly subverting the implementation of those same ideas."
>
> Gordon MacKenzie

Authentic product development innovation leadership

If you're in management, chances are that you're an unintentional roadblock to innovation. Fortunately, there are people like David Kelley who prove that innovation can be fostered, enhanced, and yes, managed. David has arguably done more for the profession of product design than perhaps any other person. He was the first to insist on credits for product engineers as well as designers in awards publications. He is the epitome of creative leadership in product development and consistently mentors others to be more creative themselves. Because he believes that design affects everything, David challenges management to lead their organizations to be more innovative. Managerial bureaucracies are generally fatal to innovation, so David has made a career of being fatal to bureaucracies. He leads by example, something that managers everywhere must come to understand and practice if they are to develop the products and positioning in the marketplace that their employers require.

TOOLS, TACTICS & TALENT

> *Manage by example, not authority*
> *Foster creativity and innovation*
> *Serve your staff rather than manage them*
> *Allow independent thinking and initiative*
> *Encourage risk-taking and failure*
> *Insulate staff from bureaucracy*

RESULTS & BENEFITS

> *Increased innovation and creative thinking*
> *More independent and responsible designers*
> *Highly motivated development staff*
> *Better designed products and systems*
> *Highly positive environment and culture*
> *Leadership development at all levels*

David Kelley: Founder/CEO, IDEO, Palo Alto, California

David is the founder and CEO of IDEO. David has been called one of the "most powerful people in Silicon Valley" and also been described as "the most sought-after design engineer this side of Thomas Edison." When David is not facilitating projects or encouraging designers and engineers, he is challenging students at Stanford University as a tenured professor in the Product Design Program, which integrates engineering courses with art, computer science, and business. David received a B.S. in electrical engineering from Carnegie-Mellon and an M.S. in product design from Stanford University.

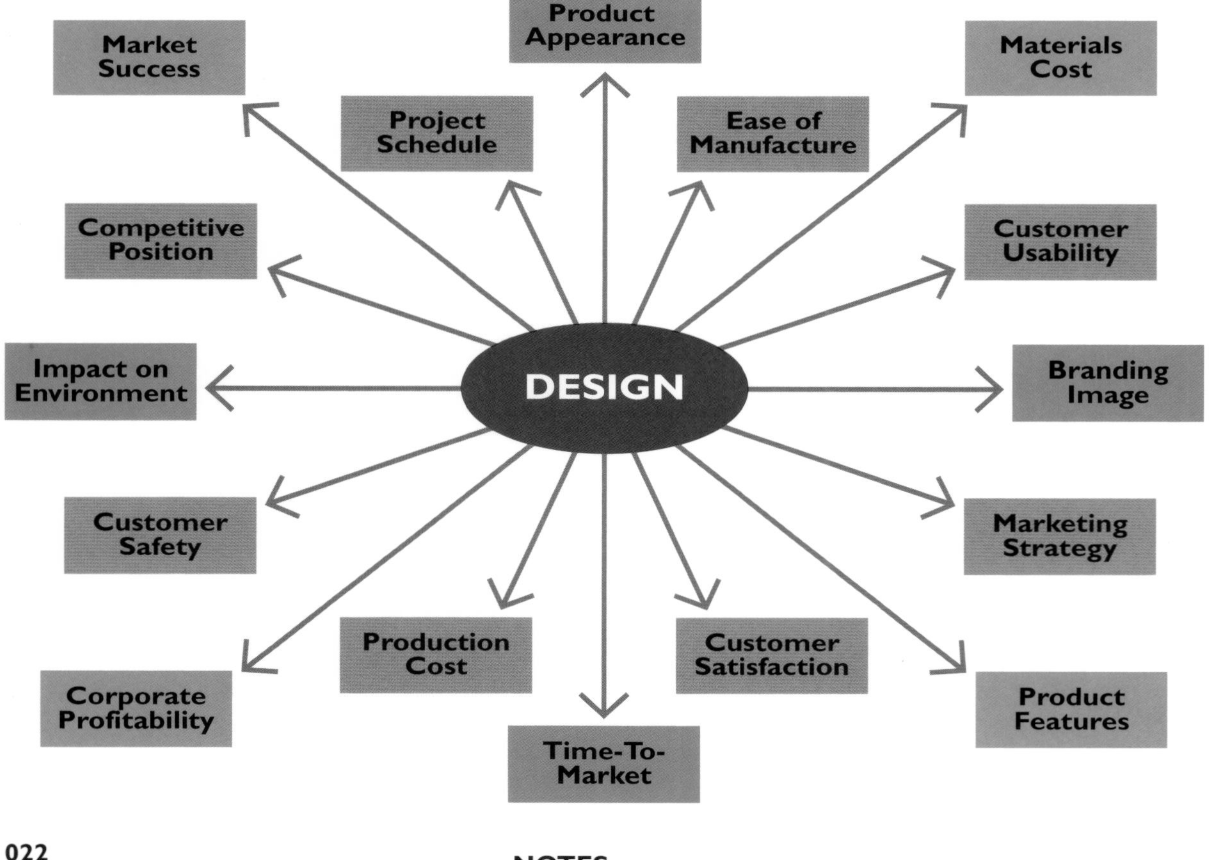

NOTES:

Terms of Engagement: Creating a successful and profitable business development model

Even though design and development consulting firms are often hotbeds of innovation and desirable work environments, it is a challenge for such firms to recruit and keep the talent required for success. The lure of benefits, stock options, IPOs, and an assortment of perks associated with the corporate world constantly tends to draw talent away from consulting firms. As new business development manager at IDEO, one of David Haygood's jobs is to help work out a business model where IDEO's creative talent is challenged professionally and rewarded financially. The diagram David sketched on a whiteboard and shown on this page depicts how IDEO works to retain some of the best designers in the world. He and other IDEO leaders constantly evaluate and refine this system.

David Haygood: New Business Development, IDEO, Palo Alto, California

David brings a broad range of business experience to his business development role at IDEO. While completing his masters degree at Stanford University in the late 1970s, he was a classmate of David Kelley and several longtime IDEO employees. Prior to joining IDEO full-time in 1999, he managed multifunctional organizations for Hexcel Composites, Specialized Bicycle Components, Raychem Corporation, and ATARI. David has managed either technical or sales organizations in industries such as automotive, recreation, aerospace, and consumer electronics.

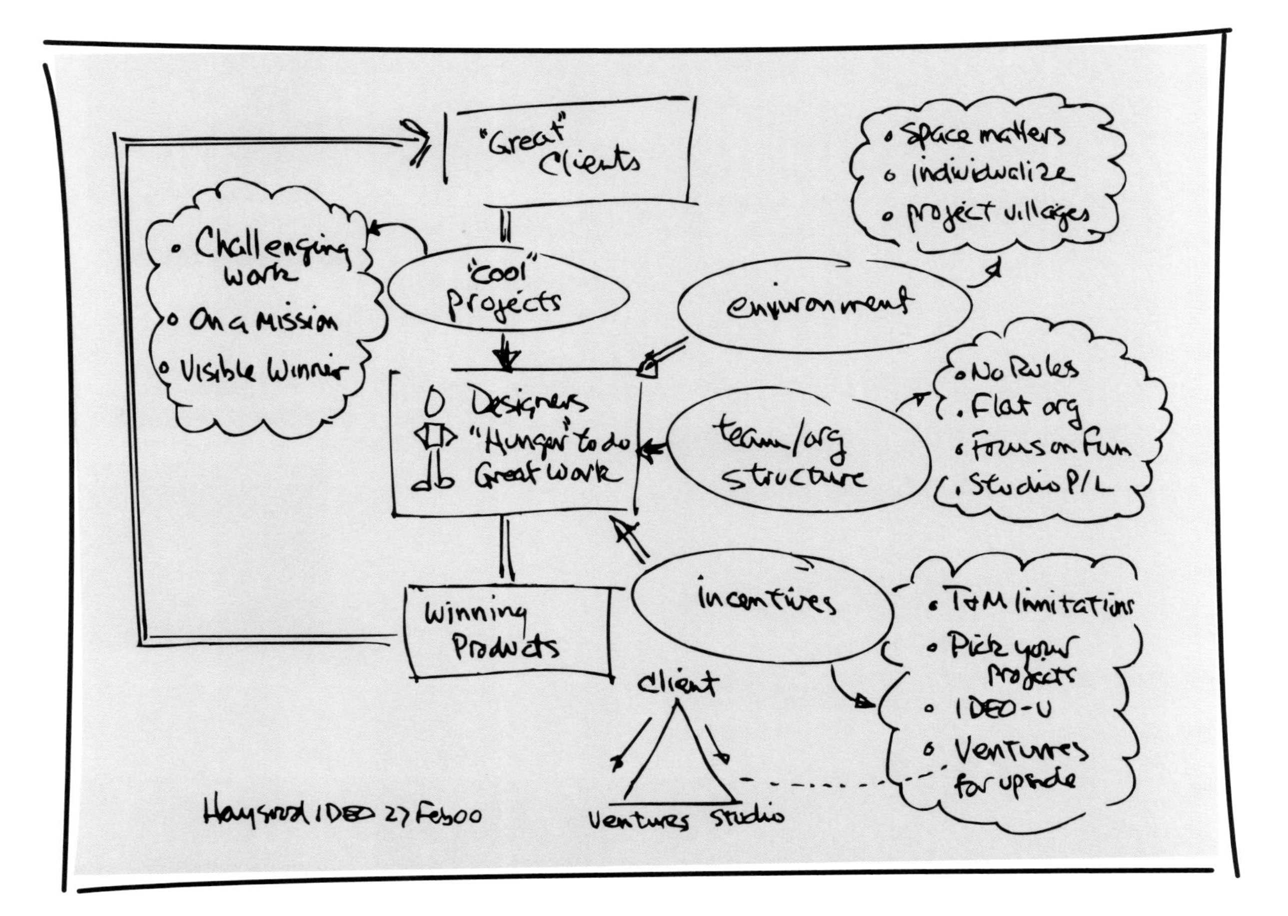

TOOLS, TACTICS & TALENT

> *Clear and concise management structure*
> *Consistency of application*
> *Opportunity for staff growth and financial benefit*
> *Treat staff as business participants*
> *Provide for employee project equity*

RESULTS & BENEFITS

> *Better retention of talented staff*
> *Entire company profits from success*
> *Staff motivated to have projects succeed*
> *High work satisfaction factor*
> *Increased innovation and motivation*
> *Ownership of projects by all participants*

NOTES:

"The person who says it cannot be done should not interrupt the person doing it."

Chinese Proverb

VIRTUAL REVOLUTION

The rapid proliferation of computer simulation, design, and development tools

Today's digital computer tools are able to accomplish phenomenal tasks that are remarkably realistic. Technology is available to scan nearly any physical object and transform it into a computer model. Processes for turning digital industrial design and engineering models into functional parts and assemblies have proliferated. It is becoming increasingly possible to conceive, design, test, and build products completely through digital means. Virtual reality tools are able to reproduce full-scale interactive digital vehicles, environments, and products. These continuously advancing processes are applicable to an ever-broadening range of applications including product development, film making, animation, human factors, entertainment, medicine, and architecture. It is imperative that the progressive enterprise take advantage of these opportunities to enhance innovation and productivity.

TOOLS, TACTICS & TALENT

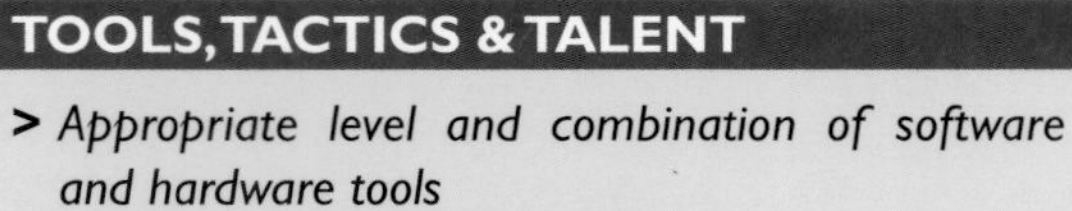

> *Appropriate level and combination of software and hardware tools*
> *Talented virtual design specialists*
> *Virtual technology champion*
> *Adequate funding and support for staff and tools*

RESULTS & BENEFITS

> *Early problem resolution via virtual development*
> *High level of visualization and understanding*
> *Highly integrated process from start to finish*
> *Optimized flexibility of designs*
> *Improved communication process*

The Revolution in Virtual/Digital/Computer Simulation and Analysis Tools

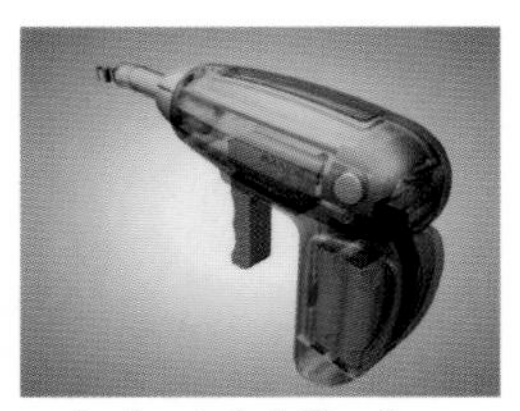
Industrial Design

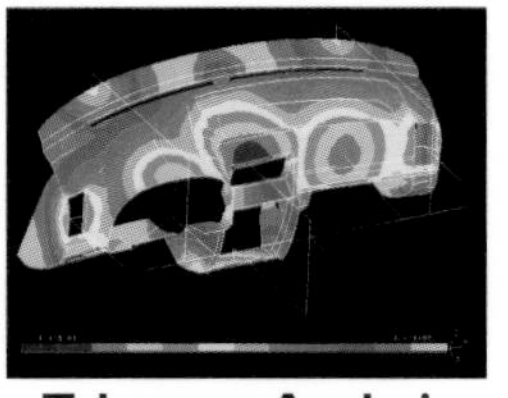
Tolerance Analysis

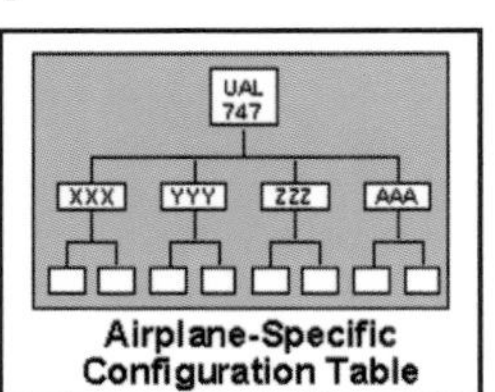

Data Management

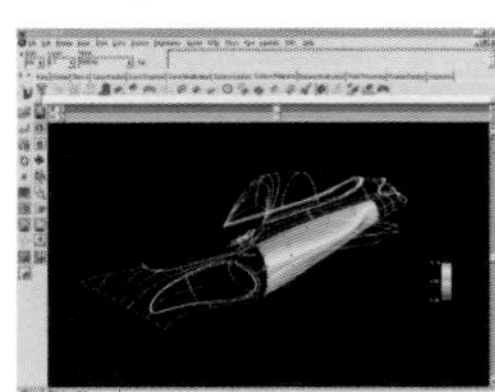
Surface Development

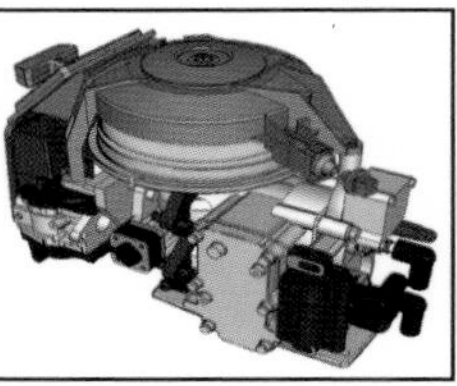
Mechanical Design

Ergonomic Analysis

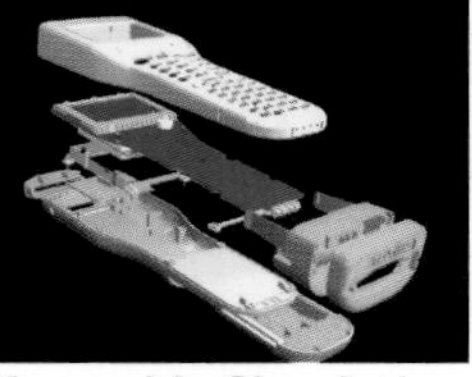
Assembly Simulation

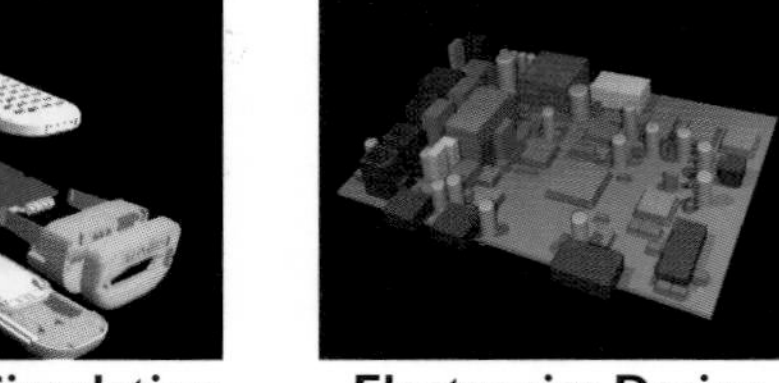
Electronics Design

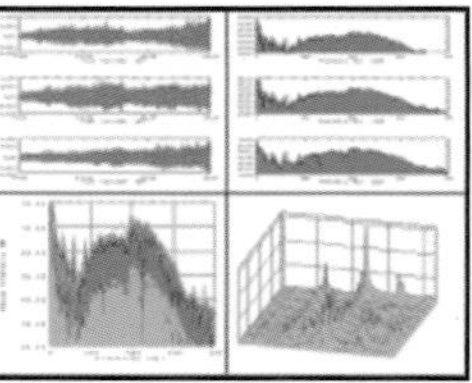
Acoustic Analysis

Digital Sketching

Blended Animation

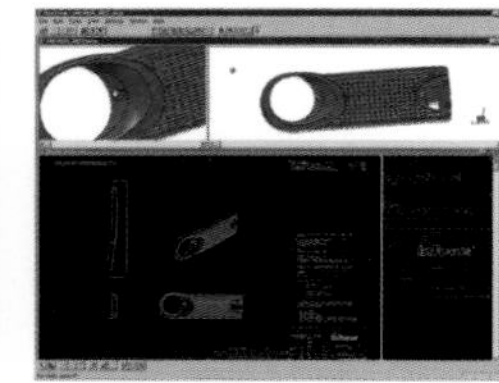
Web Interaction

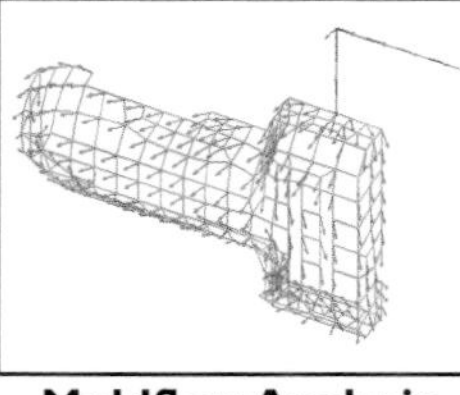
Moldflow Analysis

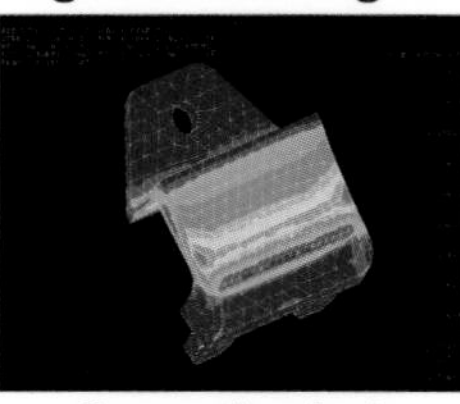
Stress Analysis

Virtual Prototyping

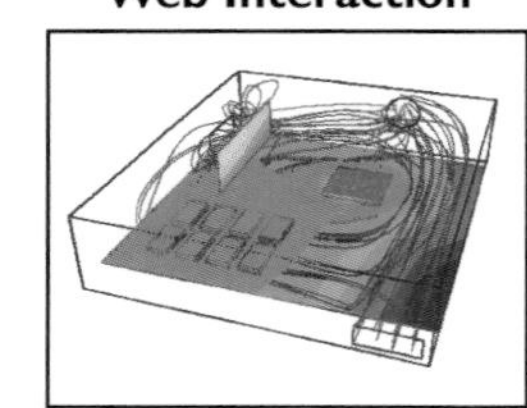
Thermal Analysis

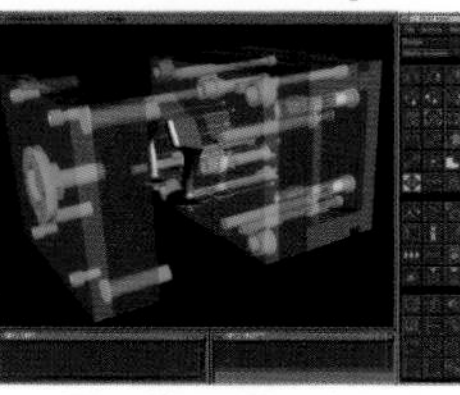
Tooling Design

Product Visualization

RPM Facilitation

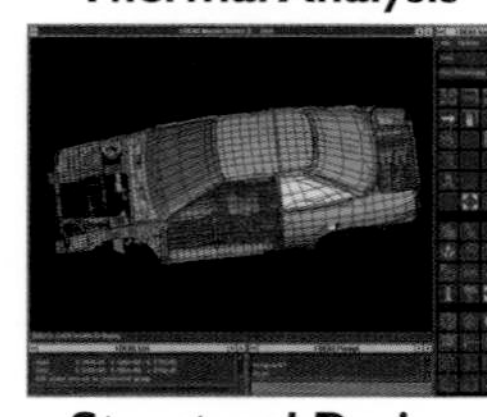
Structural Design

NOTES:

Virtual Product Development: Using digital tools to optimize critical development requirements early

Virtual product engineering is conceptualizing, specifying, designing, simulating, analyzing, testing, refining, optimizing, and validating a product primarily via digital media and models on computers. It's an iterative process that gets a design farther toward confident resolution earlier in the development process than using only the conventional, expensive, and time-consuming physical design-prototype-test-revise cycle. Much of the pain of physical shock, vibration, thermal, structural, and environmental testing in expensive labs can be minimized with early virtual simulation, evaluation, testing, and analysis done long before first physical testing in a prototype or pre-pilot run is possible (essentially too late).

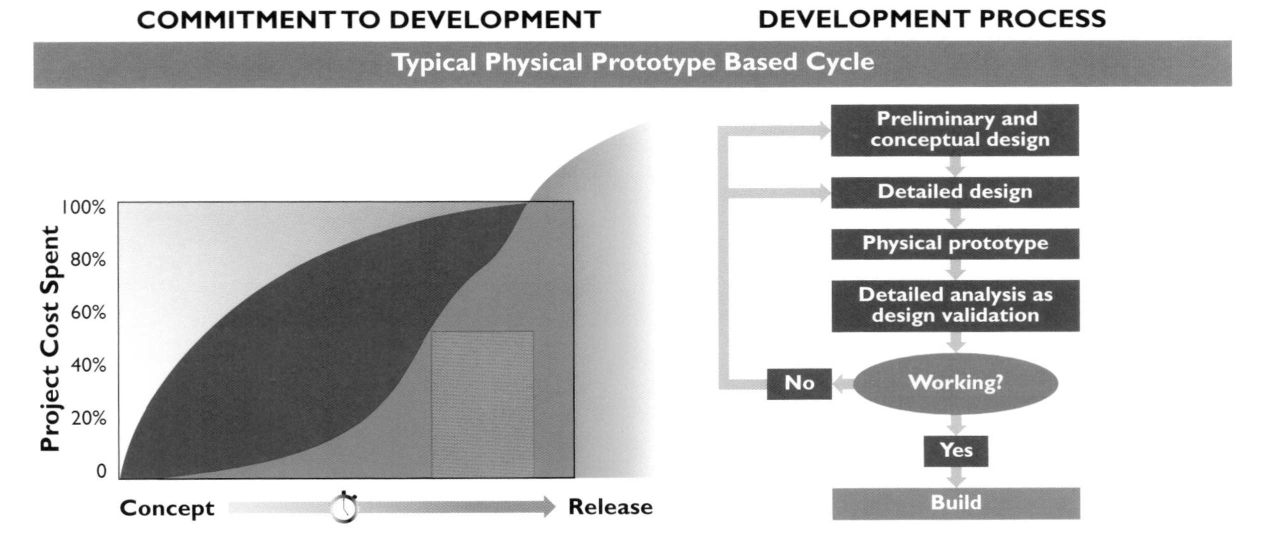

TOOLS, TACTICS & TALENT

- > *Match project spending to commitment early*
- > *Fund virtual development adequately and early*
- > *Use extensive suite of virtual development tools*
- > *Staff early with virtual design/analysis specialists*
- > *Intense virtual design-analyze-test-revise cycles*
- > *Optimize virtual design before going physical*

RESULTS & BENEFITS

- > *Early design optimization and validation*
- > *Avoidance of crash-and-burn fixes later in projects*
- > *Minimized project schedules and time-to-market*
- > *Maximized problem identification and resolution*
- > *Optimized product design features and refinement*
- > *More innovative, robust, and quality products*
- > *Reduced development, testing, and prototype costs*
- > *Maximized development ROI*

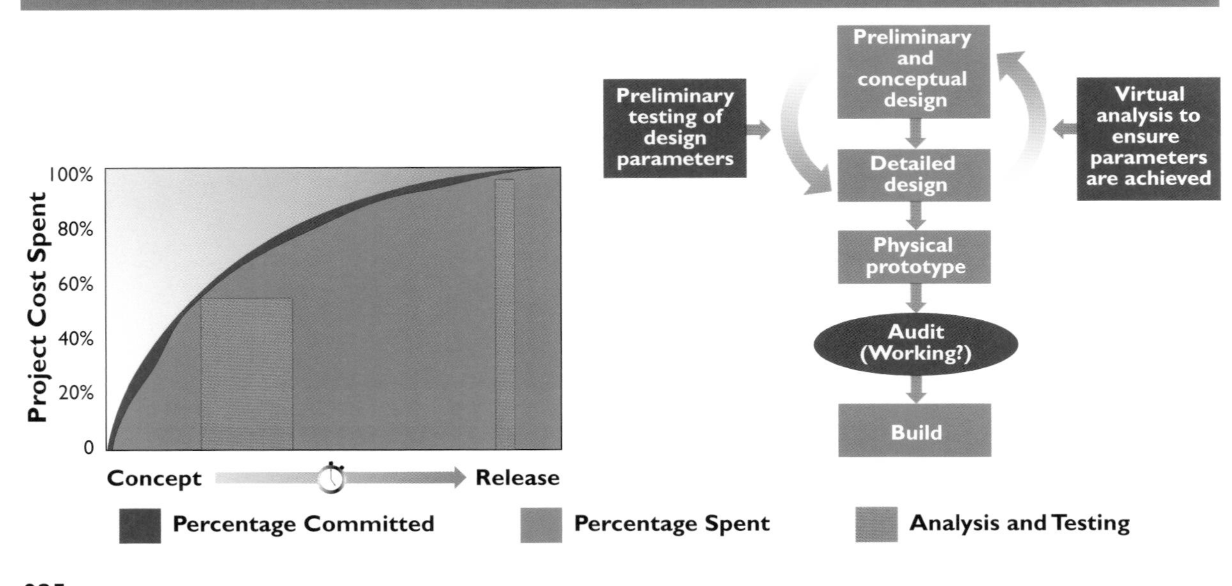

NOTES:

"An essential aspect of creativity is not being afraid to fail."
Dr. Edwin Land

SUPPORT BASED DEVELOPMENT

Providing the support resources necessary to optimize the performance of development teams

Along with the proliferation of hardware and software tools for virtual product development has come complexity that requires maintenance, training, upgrades, debugging, and other related ongoing issues. The simpler design days of pencil, paper, and T-square are long gone. It is clear that this new combination of digital tools and creative talent must be optimized. Maximum ROI is obtained by focusing a company's design talent exclusively on developing revenue-generating, innovative products and not on the maintenance of their tools. The solution is using expert specialists in as many areas as necessary to facilitate the maintenance, training, and interactive support of the development teams and their tools. Providing one support specialist for each fifteen to twenty design professionals can drastically increase productivity and performance.

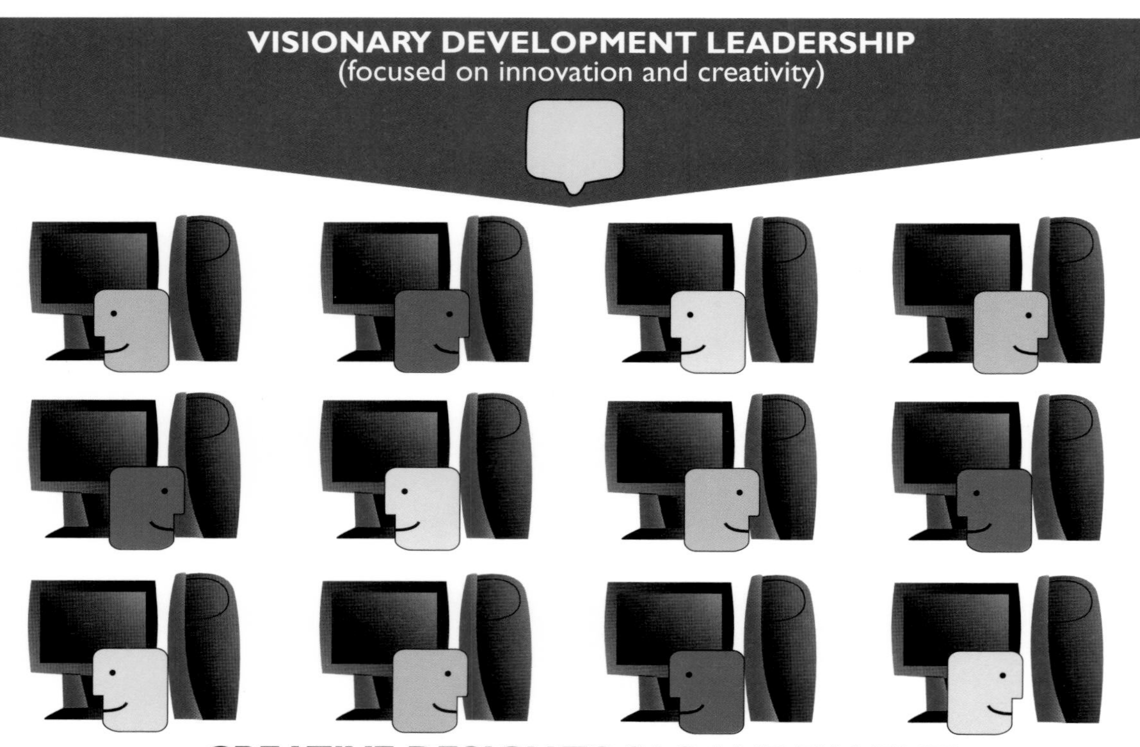

CREATIVE DESIGN TOOLS AND TALENT

TOOLS, TACTICS & TALENT

- > *Capable and talented development staff*
- > *Design talent focused on innovation and design*
- > *Experienced, capable pro-active support specialists*
- > *Best-in-class hardware and software tools*
- > *Timely access to any support services needed*
- > *Continuous training and maintenance processes*

RESULTS & BENEFITS

- > *Increased innovation and productivity of staff*
- > *Minimized frustration of development personnel*
- > *Optimized hardware and software ROI*
- > *Expanded development capabilities*
- > *Better interface with other departments*
- > *Optimized product design and development*

NOTES:

BEGIN
MGMT & FACIL
INNOV & DESIGN
MFG & OPS
BUS & MKTG
END

Support Based Success: Supporting over twenty world-class product design engineers is a full-time job

At a company like InFocus Corporation, market competition is as intense as it gets, and the creative talent of its designers and engineers must be maximized. Here, the value of support specialists is absolutely crucial to success. Design teams at InFocus are constantly pushing the envelopes of speed, quality, innovation, and integration with state-of-the-art computer design and engineering software and hardware. David Mulholland must keep up with the necessary technology and support to keep these teams running at peak efficiency. His job is to ensure that the tools, tactics, and talent of the InFocus product design teams are optimized at all times through training, software upgrades, hardware upgrades, maintenance, and one-on-one assistance.

David Mulholland: Senior CAD Applications Engineer, InFocus Corporation, Wilsonville, Oregon

David has many varied responsibilities at InFocus that include support, administration, and training for mechanical computer aided engineering software and hardware throughout the company. David is responsible for making available the appropriate information and tools for all phases of the product development process using MCAD tools. He also teaches courses on solid modeling at the Oregon Institute of Technology, presents technical seminars on product data management, and consults for a number of high-technology companies.

Critical hardware and software maintenance functions

Review of software application process

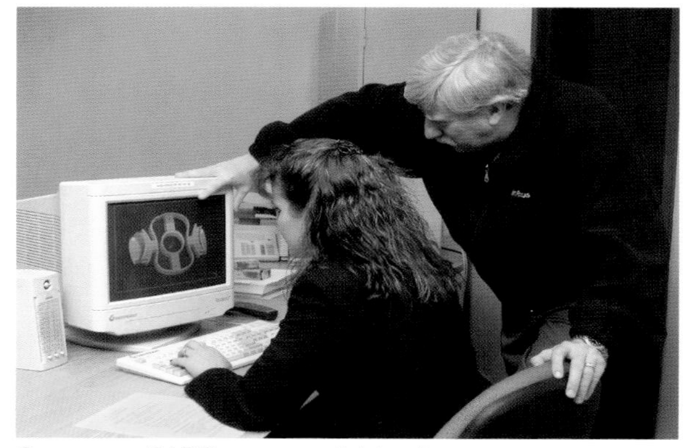

One-on-one CADD support and debugging

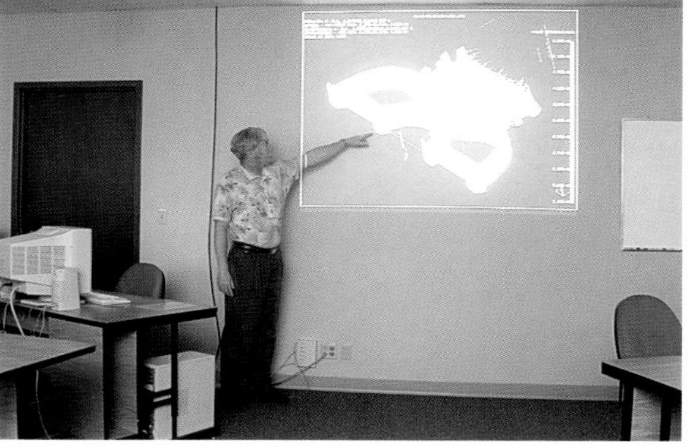

Class training session

NOTES:

"*Orville Wright did not have a pilot's license.*"
Gordon MacKenzie

DEVELOPMENT FUNCTIONS

Optimizing projects by understanding differences and strengths in physical design functions

The industrial design, product engineering, and manufacturing engineering disciplines often drive the phases and characterization of a project even though the product technology may be software or electronic in nature. It is important that management understands the various types of functional disciplines that are often lumped together into product design and engineering. Such functions can have important variations of application that can make or break the success of a project. The chart on this page shows the rough schematic position of these functions in a project as it moves from start to finish. There is obvious overlap between these functions where there are common tools, tactics, and talent as well as the important inherent handoffs of design information and data.

TOOLS, TACTICS & TALENT

- > *Mutual understanding of functional features*
- > *Clear definition of functional responsibilities*
- > *Optimized overlap of functional duties*
- > *Proper allocation of functions to phases*
- > *Appropriate training for each discipline*
- > *Systematic integration of tools and processes*

RESULTS & BENEFITS

- > *Optimized product development productivity*
- > *Appropriately allocated talent and capabilities*
- > *Adequate resourcing of project phases*
- > *Minimized conflict between functional groups*
- > *Optimize teamwork and collaboration*
- > *Better cooperation between corporate functions*

Physical Design Categories and Functions

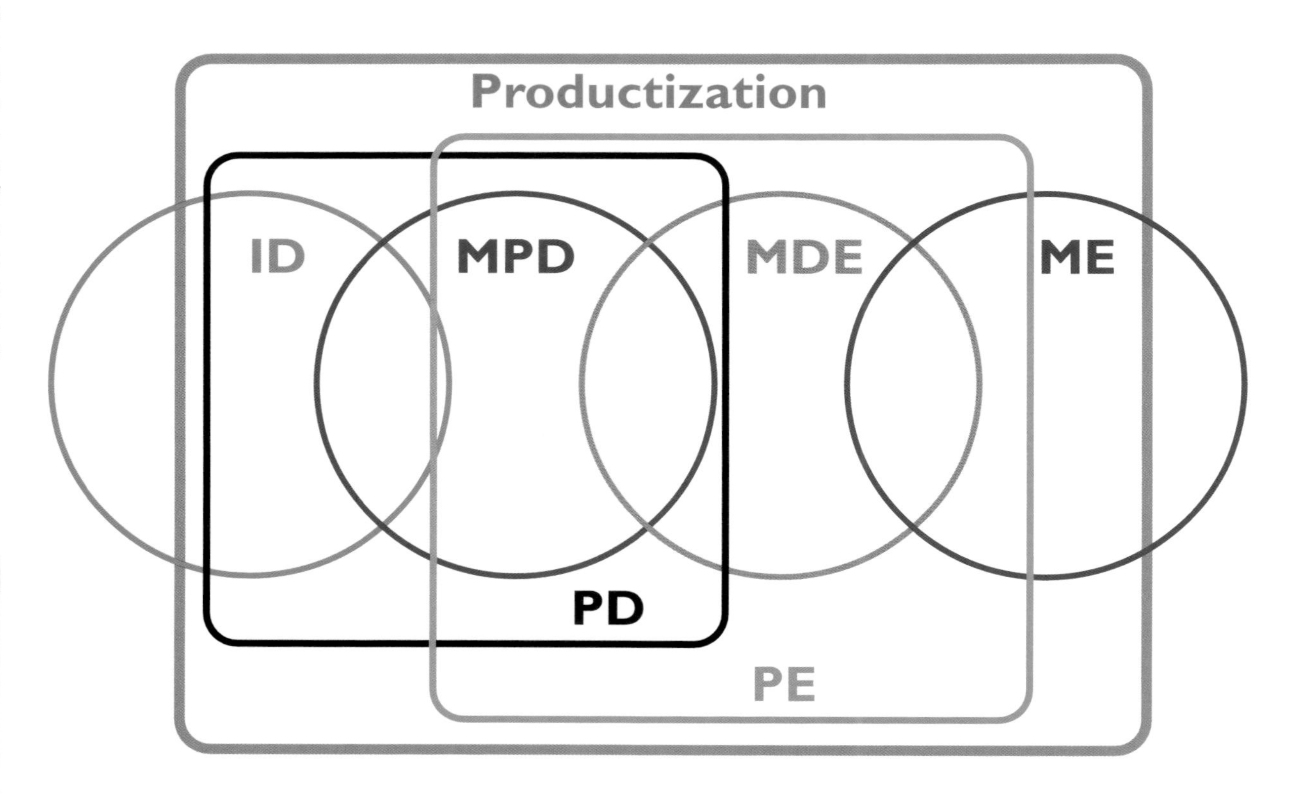

LEGEND

ID = Industrial Design: product form/appearance/styling, human factors, user interface, product language
MPD = Mechanical Package Design: enclosure, product configuration/architecture, electromechanical
MDE = Mechanical Design and Engineering: mechanical, mechanism, structural, analysis, thermal, noise
ME = Manufacturing Engineering: DFM, DFA, DFS, production, fixtures, tooling, secondary operations
PD = Product Design: mechanical package design and industrial design combination
PE = Product Engineering: mechanical, electromechanical, thermal, structural, optical, DFM, DFA

Productization = taking a viable technology and turning it into a human-usable product

NOTES:

Productization vs. Technology: Making certain that both processes are properly implemented

Two processes in that are often lumped together—and managed in the same manner, during the same time frame—are technology development and the productization of technology. These activities are quite different in many aspects and must be managed, timed, sequenced, and resourced differently. The application will depend on the intensity of each function in any particular project or product. Some products, like simple toys, have low technology development and high productization factors, whereas a digital/video projector has both high levels of technology development and productization factors. An example of a product with low productization and high technology development factors is an OEM disk drive component. The diagrams on this page show some of the characteristics of each of these processes and their integration requirements.

TOOLS, TACTICS & TALENT

> *Understanding of productization and technology differences*
> *Allocation of appropriate staff to each process*
> *Proper sequencing of technology and productization*
> *Appropriate transition of technology to productization*

RESULTS & BENEFITS

> *Optimized sequencing and flow of entire projects*
> *Minimized project problems, errors, and conflicts*
> *Shortest possible project schedules*
> *Smooth transition from engineering to manufacturing*
> *Enhanced productization of a technology*

PRODUCTIZATION

Product architecture
Component configuration
Industrial design
Enclosure design
Mechanical design
Thermal/noise management
Interaction design
Design for manufacturing
Ergonomics

TECHNOLOGY DEVELOPMENT

Physics
Chemistry
Optics
Electronics
Performance
Robustness
Reproducibility
Precision
Reliability

PRODUCTIZATION vs. TECHNOLOGY DEVELOPMENT OPTIONS

CONCURRENT: **Productization started too early before technology is stable**

TECHNOLOGY DEVELOPMENT
PRODUCTIZATION

RESULTS:
Bad process
Reduced quality
Long lead time
Painful work
Higher costs
Late to market

SERIAL: **Productization started too late after technology is locked**

TECHNOLOGY DEVELOPMENT
PRODUCTIZATION

RESULTS:
Bad process
Reduced quality
Long lead time
Painful work
Higher costs
Late to market

STAGGERED: **Both processes optimally phased for proper integration**

TECHNOLOGY DEVELOPMENT
PRODUCTIZATION

RESULTS:
Efficient process
Best quality
Shorter time
Positive work
Lower costs
Integration

NOTES:

SIX PHASES OF A PROJECT

1 - Enthusiasm
2 - Disillusionment
3 - Panic
4 - Search for the guilty
5 - Punishment of the innocent
6 - Praise and honors for the non-participants

"Collaboration is like romance. It can't be routine or predictable."

Michael Shrage in *Serious Play*

INNOVATION & DESIGN

INNOVATION FILTER-FUNNEL

An iterative development process that refines multiple product ideas down to one revenue product

The concept of iterative cycling of ideas to flesh out the best is not new, but applying numbers to it takes the process a step further. Taking three steps from twelve initial ideas to one viable product, where each time the survivors are cut in half, is quite qualitative and subjective. But providing numerical substance gives a useful paradigm to work from. Obviously, more ideas means higher chances of ending up with one or more viable, profitable, and innovative products. The antithesis of this is starting with a few pet ideas and driving them exclusively, wasting money and effort and leading to company failure. Contrary to some thinking, ideas can't hurt you, but there's nothing like a bad idea brought to market for killing an enterprise. Spending the time and resources up front on inexpensive ideation pays off in the long run.

TOOLS, TACTICS & TALENT

- > *Many brainstorming and ideation sessions*
- > *Lots of visualization and simulation processes*
- > *Involvement of cross-functional disciplines*
- > *Industrial designers and product engineers*
- > *Operations and manufacturing people*
- > *Product marketing managers*
- > *Visualization and facilitation specialists*
- > *Customers and users*

RESULTS & BENEFITS

- > *Higher probability of success*
- > *Increased product innovation*
- > *Early teamwork and collaborative interaction*
- > *Visibility of ideas and processes*
- > *Intellectual property bank of many ideas*
- > *Increased enthusiasm, motivation, and involvement*

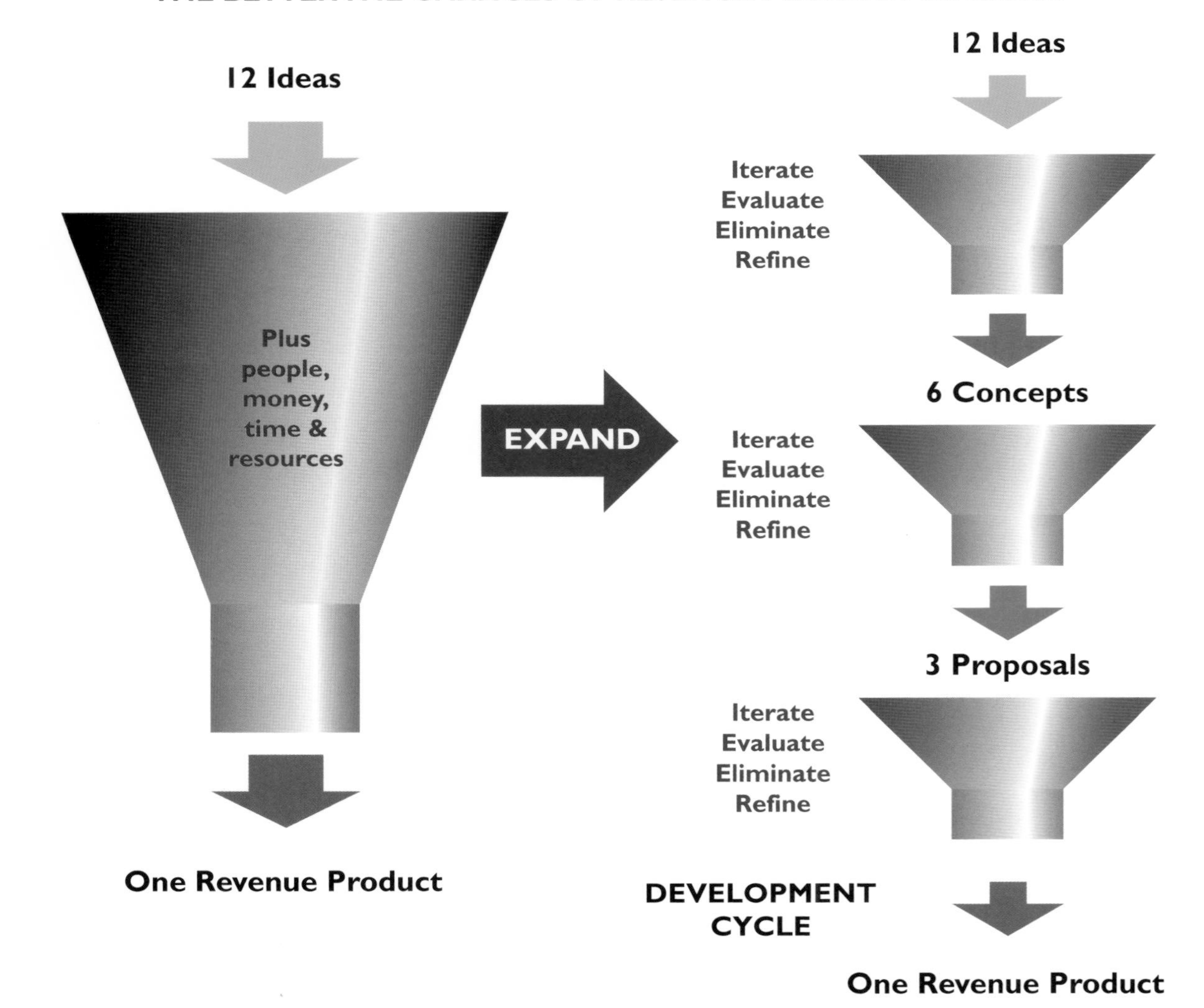

NOTES:

Tech Box System: A creative idea resource for product design innovation

It is difficult to generate ideas in a vacuum. One needs inspiration, stimuli, food for thought. Dennis Boyle knew this years ago when he started filling a card-board box with interesting materials, gadgets, and components to help stimulate creative ideas. This early personal tool of Denny's has expanded to become the IDEO Tech Box system implemented at all company locations. It is a collection of technological artifacts that anyone can look at, touch, play with, or explore. The materials in the Tech Box have been developed for other applications, but by making these stimulating items available to creative thinkers, new applications are often triggered. The outcome is more innovative product design. Designers at IDEO are encouraged to frequently peruse Tech Box materials, as well as contribute to the supply of interesting technology. Denny says, "Using the Tech Box is akin to children's creative and explorative play with a box of toys. It's the Montessori Method for product design inspiration and research."

TOOLS, TACTICS & TALENT

- > *Adequate location conducive to interaction*
- > *Accessible storage for materials and objects*
- > *Facilitate and encourage frequent usage*
- > *System curator and creative champion*
- > *Computer catalog of items and information*
- > *Constant recycling/resupply of contents*

RESULTS & BENEFITS

- > *Team access to latest design technologies*
- > *Increased team creativity and interaction*
- > *Serendipitous "ah-ha" innovation events*
- > *Increased invention and intellectual property*
- > *Breaking design roadblocks and stagnation*
- > *Improved product innovation*

Dennis Boyle: Engineering Studio Leader, IDEO, Palo Alto, California

Denny graduated from Stanford University with an M.S. in product design in 1979 and continues to teach at the Stanford Design Division as a consulting professor in product design. He is an extremely creative and innovative individual and an outstanding product designer and mechanical engineer. Denny has worked with David Kelley and IDEO for twenty-one years, contributing to literally hundreds of world-class designs and products. He holds a number of patents and has received numerous awards and recognition for his design contributions.

Tech Box System (sample items at right)

Gelastic material

Unique candy

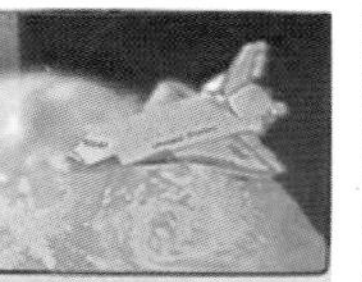

Hologram image

LED lamp

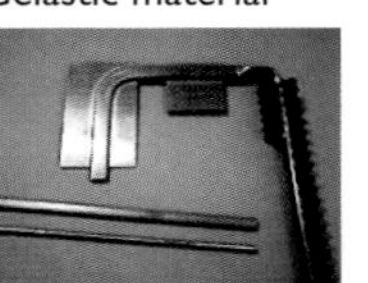

Heat pipes

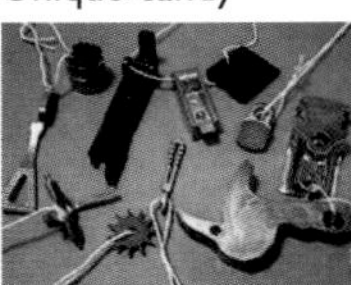

Molded parts

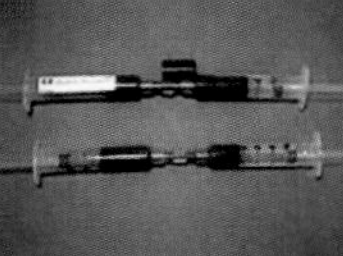

Fluid latch system

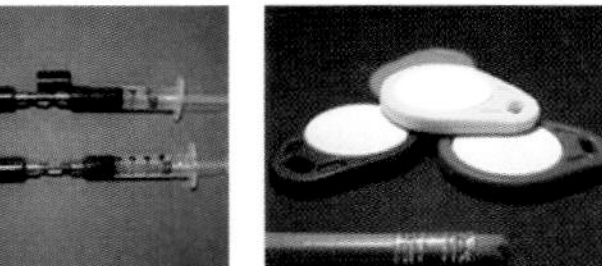

RF ID tag system

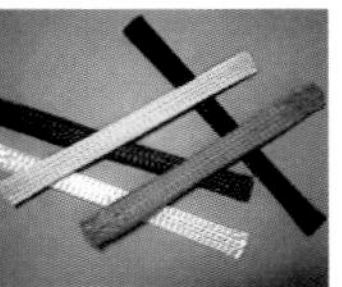

Special braid

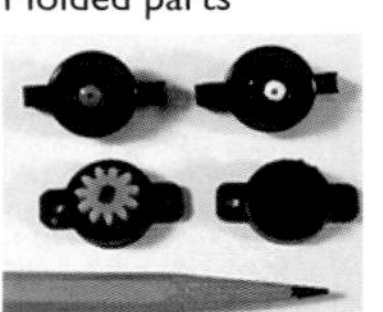

Tiny fluid dampers

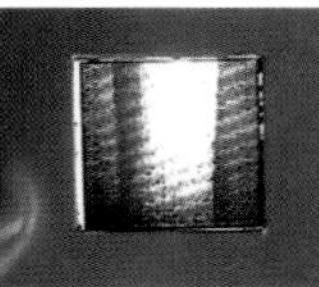

Prototyping sample

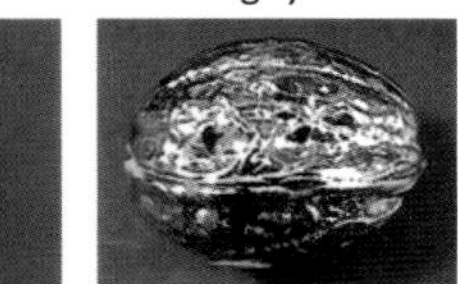

Plated walnut

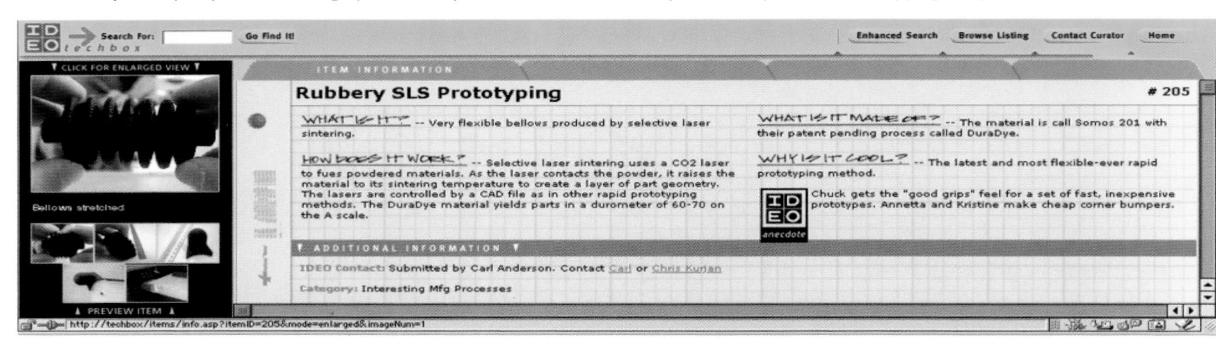

Tech Box computer database screen shot

NOTES:

EARLY IDEATION PROCESS

"Visionary speeches and mission statements embodying consultant-certified strategies don't cut it."

Michael Schrage

Essential processes using tools and tactics that will insure innovative ideas early

Essential to any development program is the early ideation process. This phase is often neglected, avoided, ignored, or, at best, poorly executed. Early ideation is arguably the most important phase of product development. It is where ideas and concepts are profusely generated and acted upon. It is the time when everyone can appropriately and freely communicate their ideas and get them out for evaluation. Executing this phase both early and extensively helps avoid having a good idea or a serious problem show up later when implementation or resolution would be less effective or detrimental. Important to the early ideation process is the use of a multitude of tools, tactics, and talent that optimize the process. The ideation process can be applied successfully anywhere across the enterprise early in any effort that requires innovation and creativity.

TOOLS, TACTICS & TALENT

> *Ideation is the first phase of any project*
> *Start early and continue throughout a project*
> *Interact and collaborate with many*
> *Employ cross-functional involvement*
> *Start free and open, and focus later*
> *Iterate often and build on all ideas*

RESULTS & BENEFITS

> *Vastly improved innovation and creative ideas*
> *Expanded ideas for other applications*
> *Early identification of problems and issues*
> *Least expensive problem-resolution period*
> *Team-building and collaboration of departments*
> *Avoidance of later problems or missed solutions*
> *Better and more profitable products*

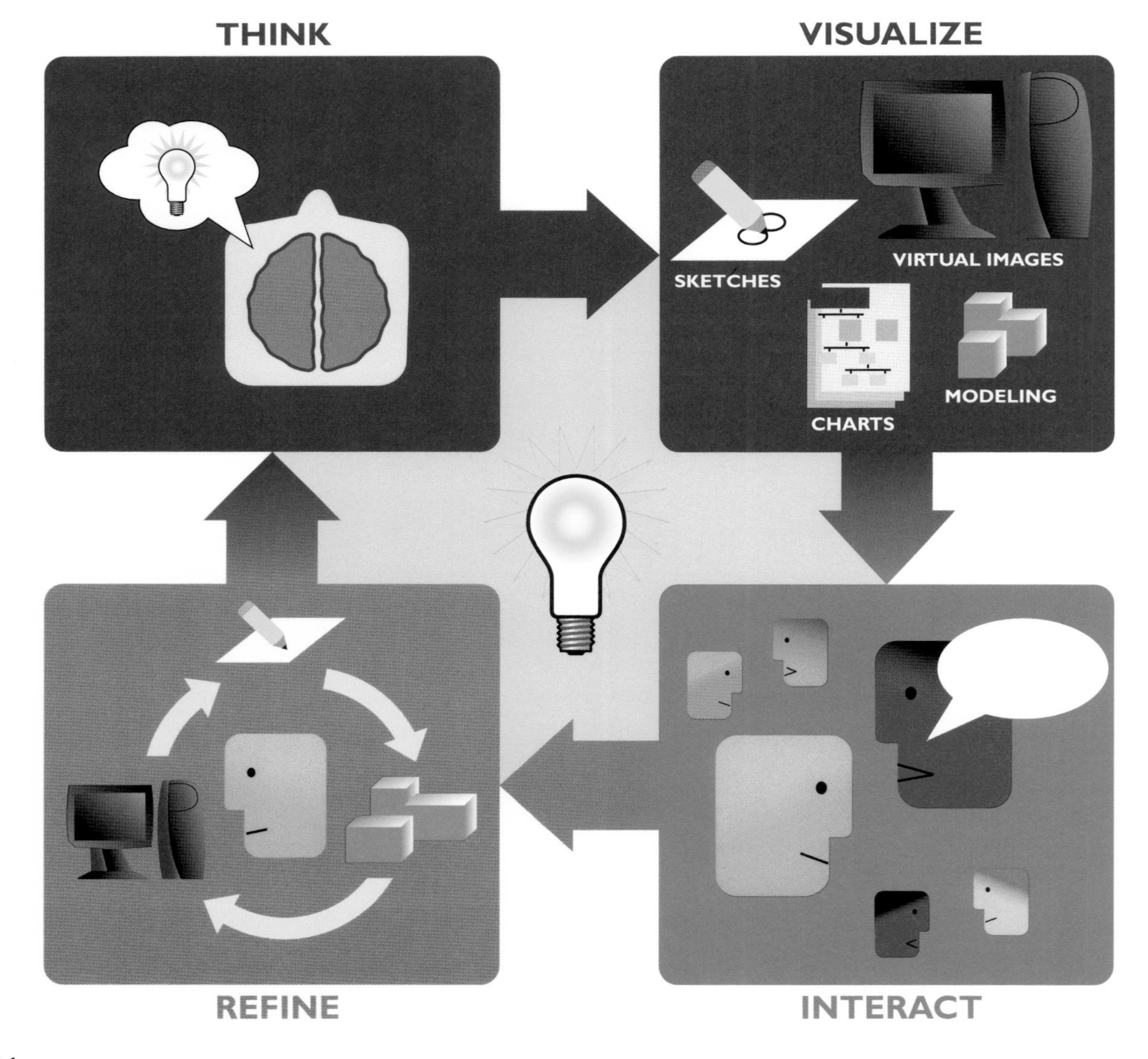

NOTES:

Breaking Innovation Barriers: Tools and techniques for overcoming blocks to creativity

There are many tools and tactics available to the development innovator for generating creative ideas and breaking down the barriers to innovation. In the ideation process, one's out-of-the-box thinking is often stifled due to conventional avenues of thought. The most important step is to apply techniques and tools without limitation and simply see what happens. Invariably, by using methodologies such as those on this page, new ideas and refinements will naturally emerge. It is especially important that these approaches are fostered and supported as often as possible in any corporate development environment from the early stages of a project through its completion.

James L. Adams: Author and Professor Emeritus, Stanford University, Stanford, California

Jim Adams received his B.S. in mechanical engineering from the California Institute of Technology in 1955 and also studied art at UCLA. He received both his M.S. and Ph.D. in mechanical engineering with an art minor from Stanford University. Jim is interested in how technical organizations manage innovation, and he promotes creativity and innovation through the seminars and workshops he conducts. In addition to publishing several books, he was awarded the H. Geggenheimer Award for Innovation in 1997.

Metaphors

Sticky Notes

Bug Lists

Environment

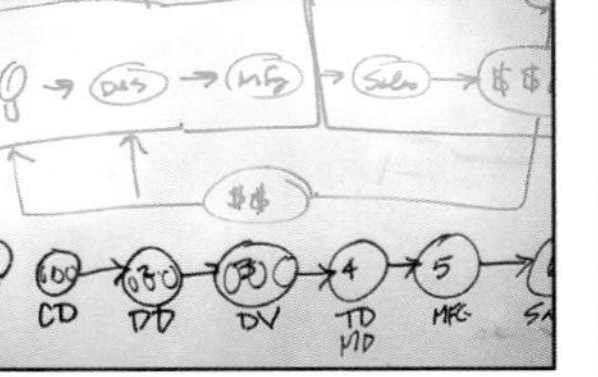
Flow Charts

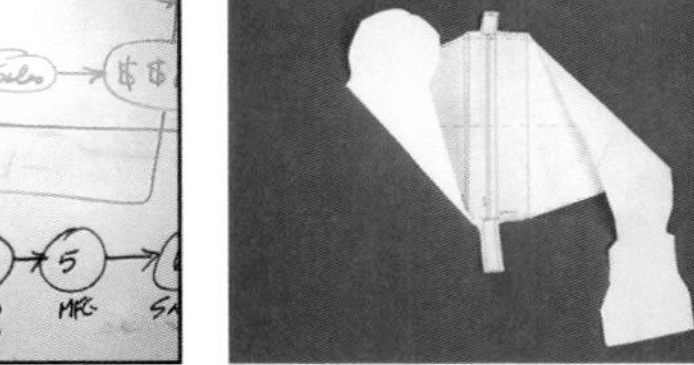
Paper Dolls

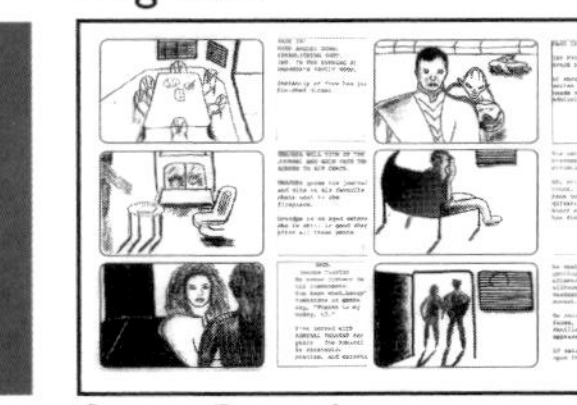
Story Boards

Role Playing

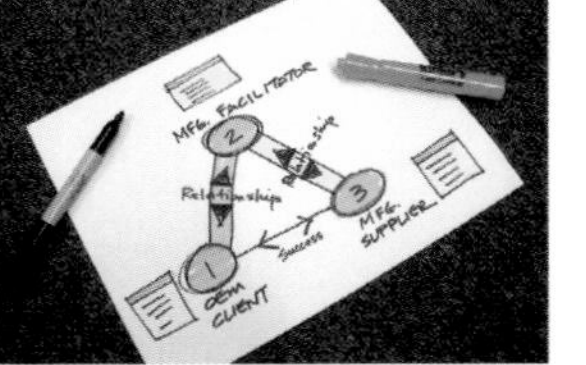
Diagrams

Sketches

Brainstorming

Cardboard

Foam Blocks

Facilitation

Directed Fantasy

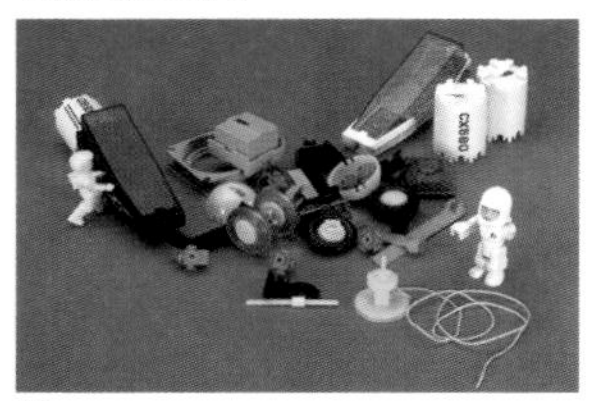
Toys

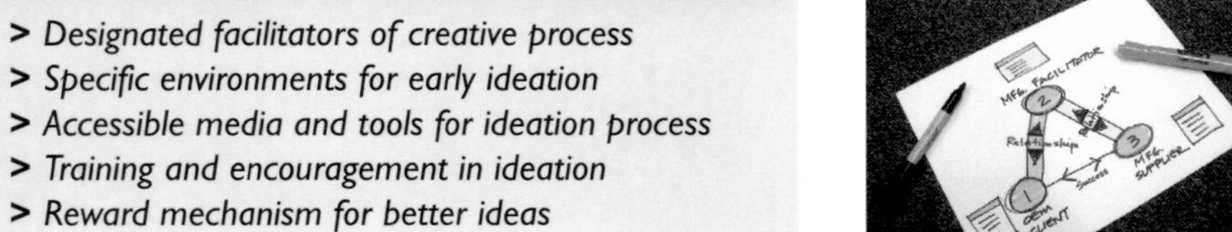

TOOLS, TACTICS & TALENT

- *Designated facilitators of creative process*
- *Specific environments for early ideation*
- *Accessible media and tools for ideation process*
- *Training and encouragement in ideation*
- *Reward mechanism for better ideas*

RESULTS & BENEFITS

- *Early problem definition and resolution*
- *Optimum product features and usability*
- *Reduced cost due to fewer late changes*
- *Encouragement and inspiration of others*
- *Motivation and increased awareness of staff*

NOTES:

Freehand Sketch Ideation: An essential visualization art using simple tools, techniques, and materials

Freehand sketching has been around since humans were first created. Even with the proliferation of computer visualization tools, sketching remains essential to the development process. Whether it's a rough schematic diagram on a napkin or a sophisticated detailed drawing, freehand sketching can and should be used across the enterprise for communicating ideas. The tools of execution are simple and readily available to both the novice sketcher and a professionally trained visual designer such as Mark Schoening. Mark's sketch work uses a broad variety of styles and techniques. He produces concepts and designs in many situations—wherever and whenever an idea needs tangibility. Mark's work ranges from simple rapid line drawings to refined marker sketches depicting detailed product features.

Mark Schoening: Co-Founder and Principal, FUSE, Inc., Portland, Oregon

Mark is an outstanding freehand design visualizer, and his sketch work is vital in many product development processes. Mark is a graduate of the Industrial Design Program at California State University at Long Beach. In addition to many honors and awards earned as a design student, Mark has demonstrated outstanding work during his professional career. When designing, Mark is able to combine his visual skills with an expert understanding of product engineering. His firm, FUSE, specializes in medical, consumer, and high-technology product development.

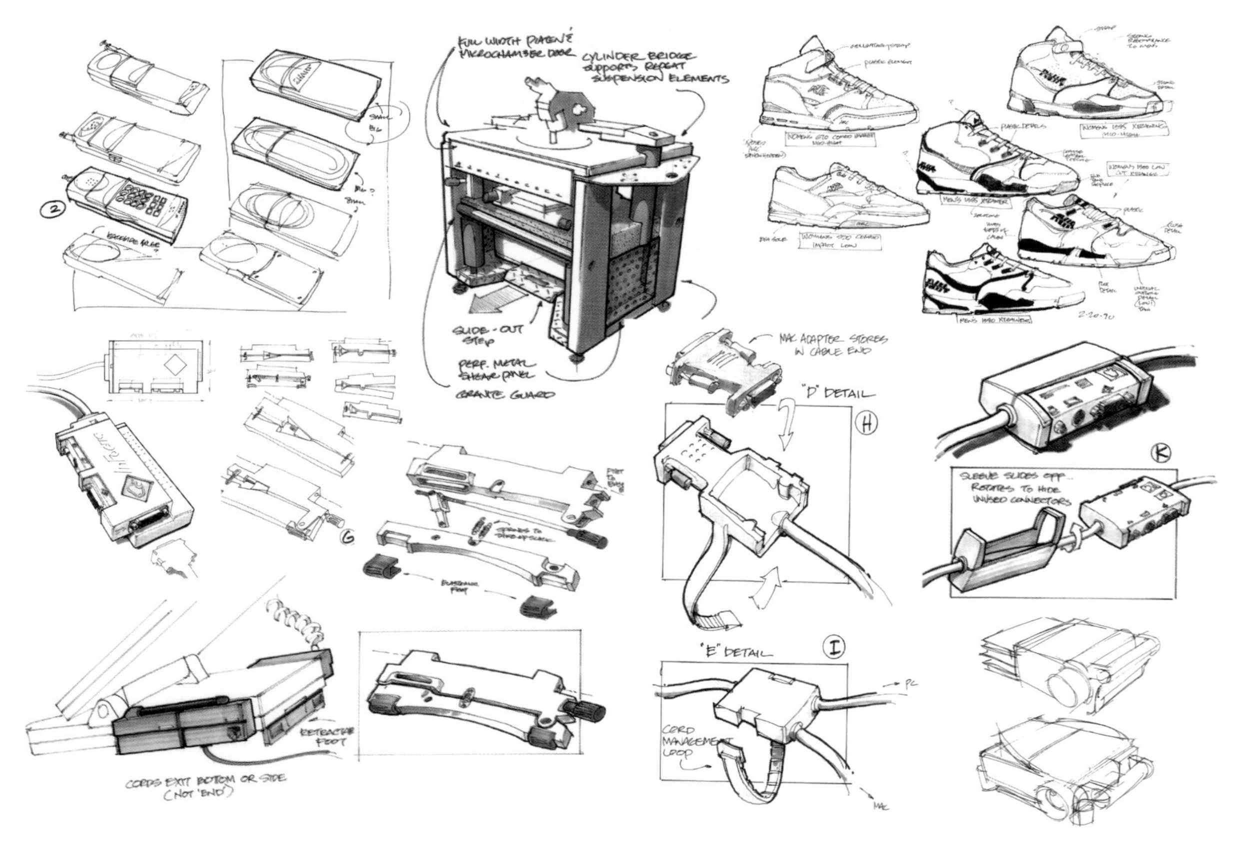

A selection of Mark's excellent design sketch work

TOOLS, TACTICS & TALENT

- *Talent pool of capable freehand visualizers*
- *Use of sketching in all areas of development*
- *Training of all designers in freehand visualization*
- *Digital conversion for archiving and expansion*
- *Encouragement of non-artists to use sketching*
- *Frequent display of results to all*

RESULTS & BENEFITS

- *Quickly executed visual ideas and concepts*
- *Simple tools and materials readily available*
- *Format that is easily digitized and reproduced*
- *Can be carried out anywhere at any time*
- *Highly interactive utilization capability*
- *Variable skill levels acceptable and viable*

NOTES:

Digital 2D Ideation: Simple tools and techniques provide quick execution with 3D realism

Virtual digital design need not require a costly or complex tool set. Steve Montgomery uses a 2D desktop computer illustration program to develop products for his clients. He has refined his techniques to a high level of sophistication and productivity. After making "thumbnail" freehand sketches of his concepts, he moves to his computer and "sketches" the product's form and graphics with the illustration software. By doing this, Steve can show variations quickly in an orthographic format that depicts several views of the product, then begin to design the product's user interface. He applies his artistic knowledge of color, form, light, and shade to give three-dimensional realism to the 2D images. This process helps him to conceptualize, illustrate, modify, and present a variety of design alternatives to optimize the final design.

TOOLS, TACTICS & TALENT

> *Desktop publishing hardware and software:*
- *digital computer and monitor*
- *flatbed scanner*
- *color inkjet printer*
- *digital illustration program*
- *digital photo program*

> *Sophisticated industrial design talent*

RESULTS & BENEFITS

> *Concepts/ideas/files easily managed*

> *Quick execution and revision capability*

> *Design options/variations explored easily*

> *Simple, inexpensive digital design tools*

> *Compact development environment*

> *Timely and productive outcomes*

> *Precise and realistic results*

Steve Montgomery: Founder and Principal, bioDesign, Pasadena, California

Steve is principal of his own product design firm, bioDesign, and teaches product design at Art Center College of Design, both in Pasadena, California. bioDesign serves a variety of clients in the California high-technology community. It also invents and develops its own products. bioDesign has recently won a Business Week IDEA *award for its design work. Articles describing Steve's work have appeared in the* Los Angeles Times *and* Business Week Magazine. *Steve is also a ballet dancer, having performed professionally for many years.*

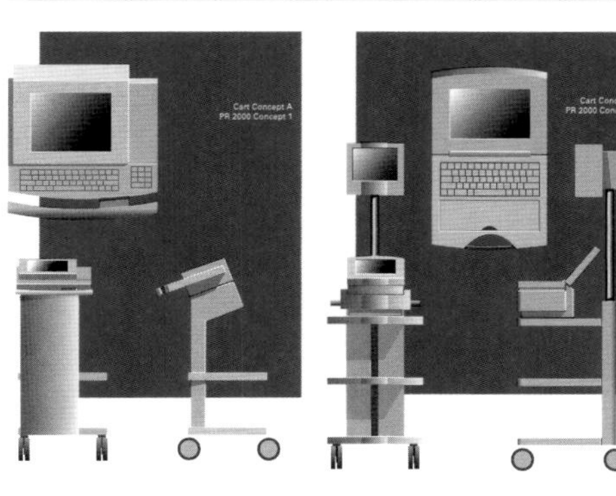

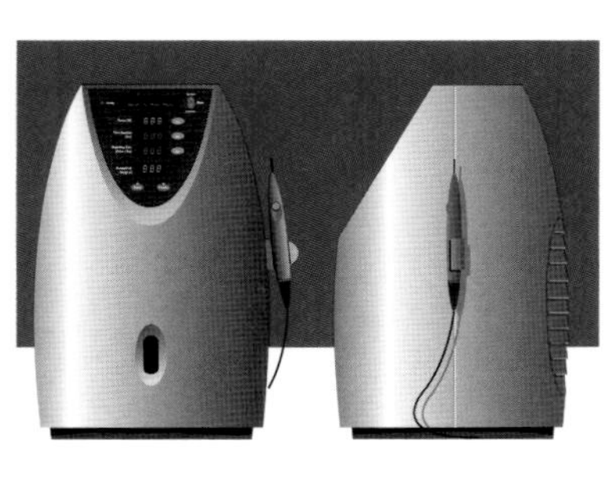

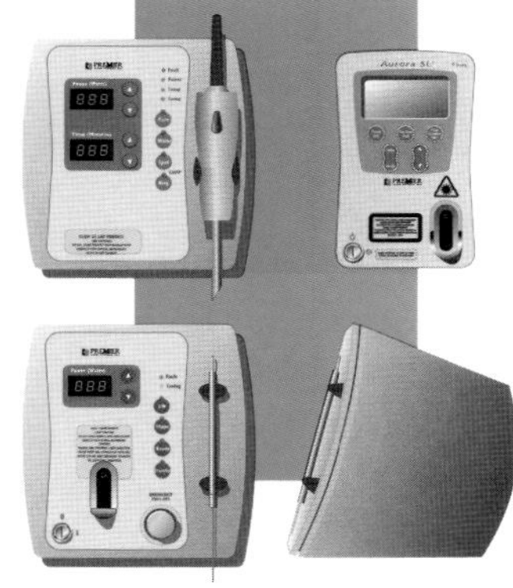

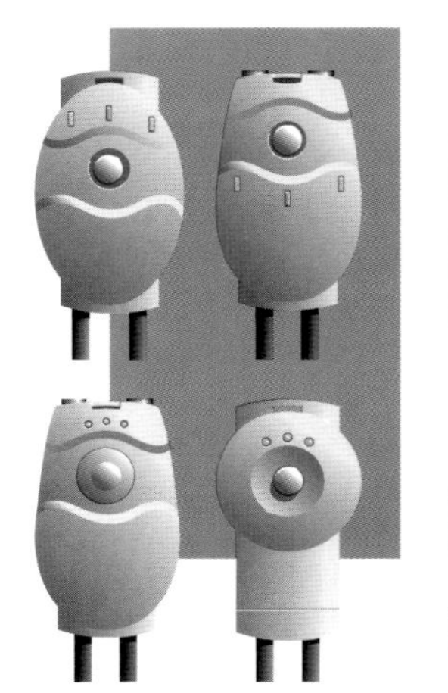

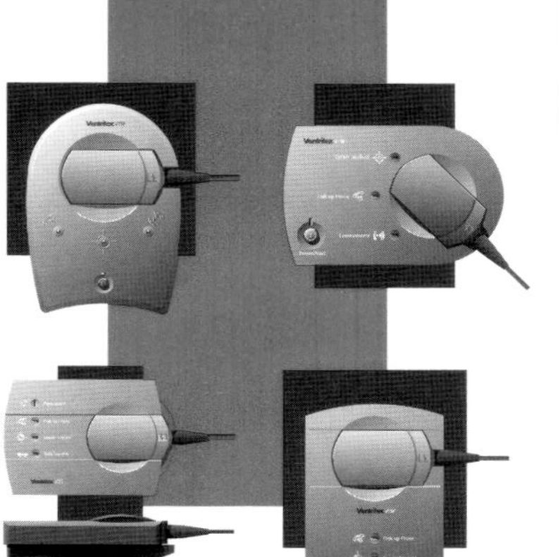

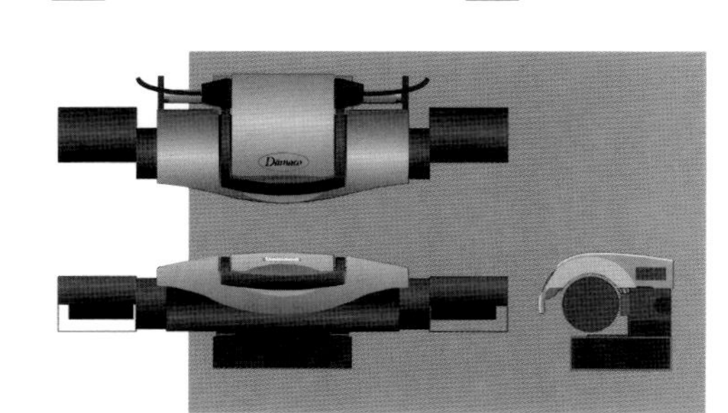

A selection of Steve's 2D digital design work

NOTES:

Digital Sketching: Transferring the art of freehand visualization to the computer

Relatively new to virtual visualization, and obviously adapted from the practice of freehand sketching with pen and paper, is digital sketching. This sophisticated process combines the art of manual drawing with the medium of the digital computer and "paint" software. A traditional computer mouse, or a combination of a pencil-shaped stylus and interactive digital sketchpad, plus a digital computer are used to expand freehand sketching into an explosion of creative imagery and new possibilities. What was once often painstaking using pen, pencil, marker, and/or airbrush can now be done digitally with ease. Digital sketching can emulate the finest manual illustration and expand into far more sophisticated techniques. These include sketching over existing 3D digital models and combining 2D and 3D imagery together, the results of which are infinitely and easily modifiable.

TOOLS, TACTICS & TALENT

- > *Talented and creative digital sketchers*
- > *Talented and creative visual designers*
- > *Appropriate software and hardware tools*
- > *State-of-the-art training and skill development*
- > *Integration of 2D and 3D images*

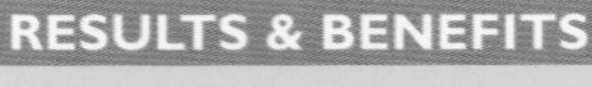

RESULTS & BENEFITS

- > *Faster image generation and revision*
- > *Improved idea communication and transfer*
- > *Multiple media image capability*
- > *Better integration of design media and process*
- > *Vastly expanded storage and retrieval of ideas*

Alexandra Walsh: Industry Marketing Manager, Alias|Wavefront, Toronto, Canada

Alexandra is the industry marketing manager for the Design Business Unit at Alias|Wavefront (A|W). Prior to joining Alias|Wavefront in 1997, she worked in a variety of marketing and project management roles. She has a B.A. in psychology from the University of Waterloo and a post-graduate diploma in public relations. Alexandra helps industrial designers, engineers, and many other design professionals discover how sophisticated software tools can optimize their visualization work. Born in Sudbury, Ontario, Alexandra now lives in Toronto.

Loose concept (Alchemy)

Refined concept (Alchemy)

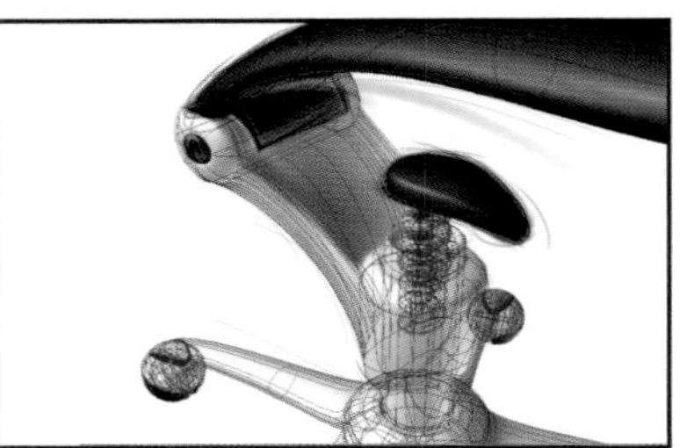

Sketching over 3D surface model (Alchemy)

White on black for effect (R. Paul, A|W)

Sophisticated coloration (D. Xiong, A|W)

Simple line and form (' 1998 Style Create Co., Ltd.)

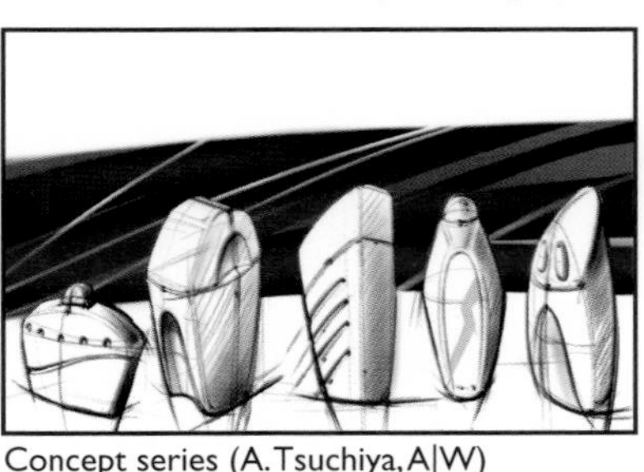

Concept series (A. Tsuchiya, A|W)

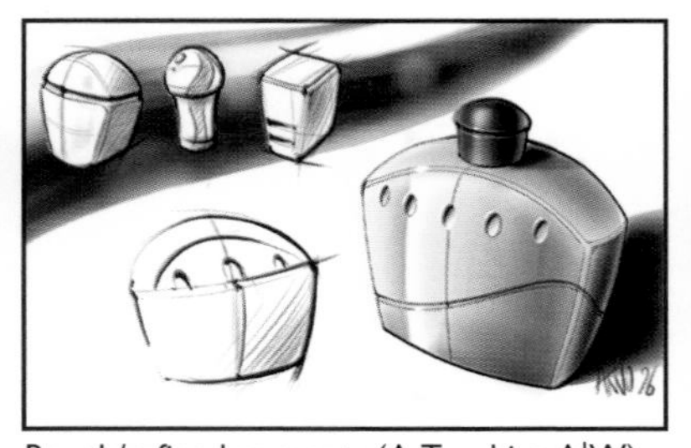

Rough/refined concepts (A. Tsuchiya, A|W)

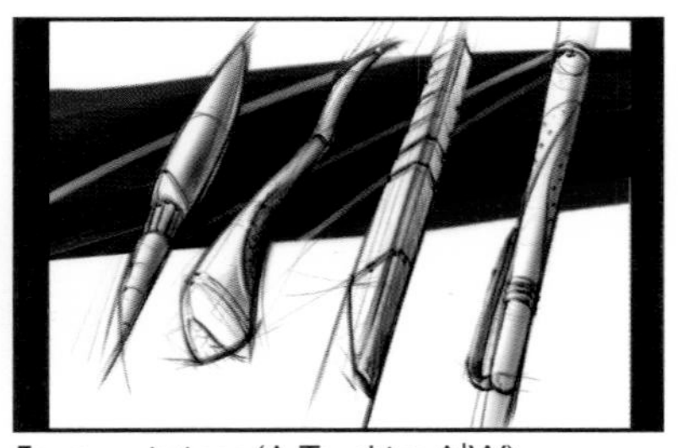

Form variations (A. Tsuchiya, A|W)

NOTES:

(All images this page courtesy of Alias|Wavefront)

Three-Dimensional Sketching: Unique manual skills beyond pen and paper

One rarely thinks of freehand sketching as a three-dimensional activity. However, in the early ideation process, it is useful to execute three-dimensional sketches with simple materials such as cardboard, paper, tape, glue, and foam. Carving knives, sandpaper, and cutting devices are the tools of expression instead of pens and pencils. Three-dimensional sketches can be crude or sophisticated, simple or complex, quickly executed or highly crafted. When the idea-generating thinker interacts with a skilled 3D sketch artist their collaboration can yield highly creative results. Herb Weiland is both an idea-generator and a highly skilled 3D sketch artist who creates quickly executed mockups as well as refined finished models. Just as in 2D sketching, much of Herb's 3D work is done with manual modeling and a minimal use of power tools.

Herb Weiland: Founder/Principal, weilandesign, San Pedro, California

Herb is principal of weilandesign, a product development consultancy that specializes in rapid 3D visualization of consumer products, primarily in the toy industry. He enables his clients to see products in physical form, after the sketches but before the final hard models are fabricated. Herb has consulted for many leading toy companies, including Fisher-Price, Galoob, Hasbro, Kenner, Mattel, and Wham-O. He received his B.S. in industrial design from Art Center College of Design, and is now venturing into fine art sculpting.

Functional structural toy mockup

Architectural space model

Master model for casting pattern

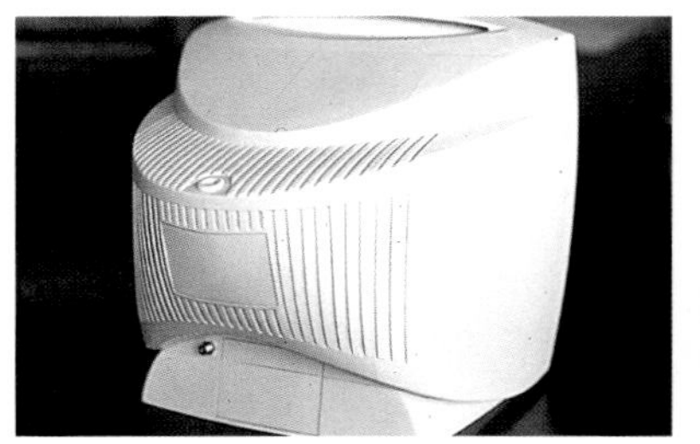

Foam mockup of consumer television

Foam rendition of helmet concept

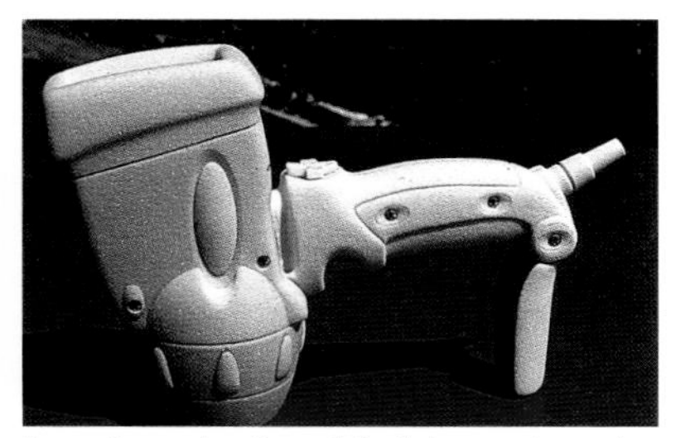

Foam form detail model of electronic tool

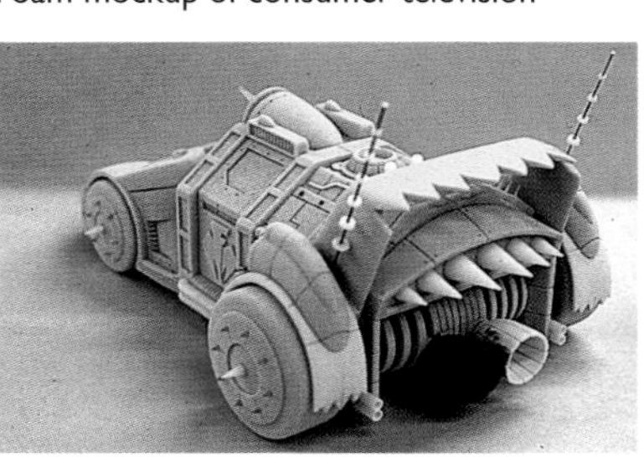

Detail foam model of toy vehicle

Detail foam model of toy plane

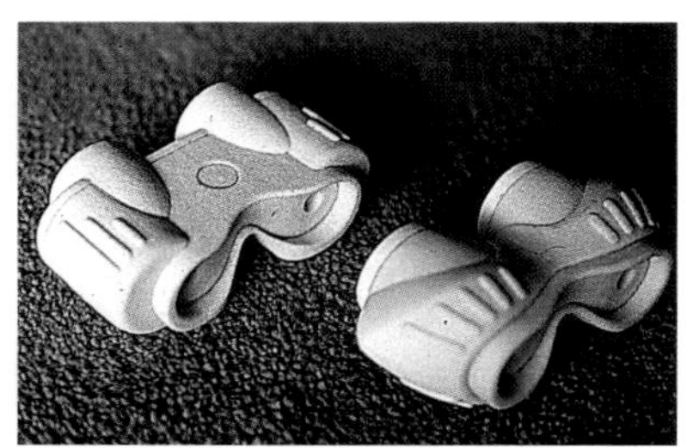

Concept mockups of children's binoculars

TOOLS, TACTICS & TALENT

- > *Talented and capable 3D visualizers*
- > *Variety of materials and processes at hand*
- > *Appropriate shops and tools available*
- > *Encouragement of staff to think and sketch in 3D*
- > *Interactive group 3D sketch sessions*

RESULTS & BENEFITS

- > *Tactile, kinesthetic approach to product design*
- > *Ability to engage others more easily than with 2D*
- > *Testability with users and customers*
- > *Improved perception of product look and feel*
- > *Capability for integrated reverse engineering*
- > *Capability to turn into 3D digital model*

NOTES:

Full-Scale Physical Simulations: There's nothing like having the real thing to interact with

Large products such as trains, automobiles, and MRI scanners have special design scale factors. Even with the advent of advanced computer simulation technologies, there is often the need for actual physical representations. Ultimately, only a full-size mockup will allow certain evaluations of human interaction, kinesthetics, look and feel, space, ergonomics, form, surface, and other design aspects. Designworks/USA has such a need for constructing physical simulations, and they do it on a routine basis. A major portion of their business is the design and development of transportation vehicles, systems, and components. Their facilities include large workshops specifically developed and outfitted to execute full-scale detailed mockups in clay, foam, wood, and other materials as appropriate to a specific project.

TOOLS, TACTICS & TALENT

- > *Adequate space and shops for construction*
- > *Appropriate tools and materials for mockups*
- > *Competent specialists in building large mockups*
- > *Proper evaluation and presentation areas*
- > *Feature, image, and surface capture tools*

RESULTS & BENEFITS

- > *Proper human factor evaluation and design*
- > *Coverage of subtle design details*
- > *Proper kinesthetic and space evaluation*
- > *User/customer interaction and test opportunity*
- > *Improved final design execution*

Henrik Fisker: President and CEO, Designworks/USA (BMW), Newbury Park, California

Henrik's first position as a designer was at BMW Technik in its Advanced Concept Studio for automotive design. He studied at Art Center College of Design in Pasadena, California, and the University of Switzerland. Since then, Mr. Fisker has worked on the BMW electric show car interior, the exterior of the Z07 show car for the Tokyo Auto Show in 1997, and the exterior of the Z8 Roadster production car, which was launched as the new James Bond car in 1999. He served as the Director of BMW Automotive Design for Designworks/USA from 1997 to 2000.

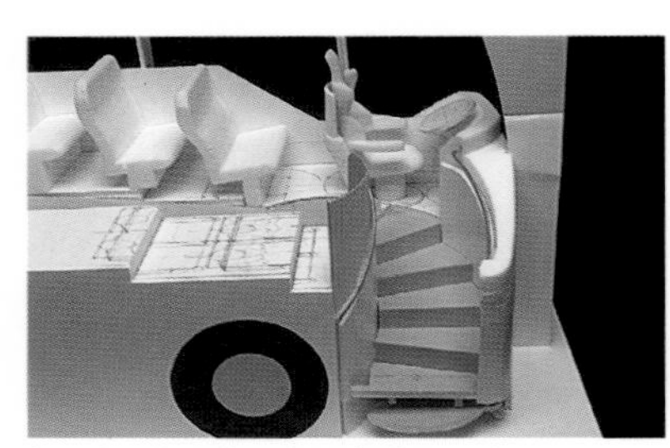

Initial reduced-scale vehicle mockup

Full-size wood mockup of bus stairway

Simulation of bus front cab area

Space mockup of train interior

Building a full-size train interior space

Evaluating full-scale facilities usage

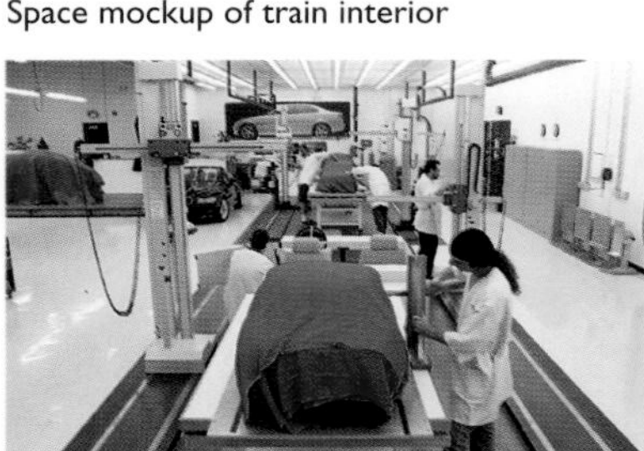

Precision clay modeling laboratory

Clay surface model of bus front section

Clay surface model of bus rear section

NOTES:

Development Serendipity: Making it happen

It is surprising how many revolutionary discoveries and inventions have been made accidentally. Many of us already know about the serendipitous innovations of penicillin and 3M's Post-it® Notes. However, there are hundreds more such discoveries that happened simply because a mistake was made, something was left to simmer by accident, or there was a casual but astute observation. A number of such serendipitous inventions are listed on this page. The real question, however, is how does one foster or "make" serendipity happen (and with beneficial discoveries) when one is trying to do something totally different? Most design solutions are outcomes of focused thought with the "a-ha" discovery resulting from engaging in good process.

"Serendipity . . . both luck and accident play a role, but hard work, alertness, and perseverance are also demanded."—Gilbert Shapiro, *A Skeleton in the Darkroom: Stories of Serendipity in Science.*

TOOLS, TACTICS & TALENT

> *Extensive experimentation*
> *Significant free exploration*
> *Environment that fosters creativity*
> *No fear of risk or failure*
> *Observe all things carefully*
> *Don't take anything for granted*
> *Simply keep trying something*

RESULTS & BENEFITS

> *Increased probability of accidental discovery*
> *More ideas to feed into the development process*
> *Higher chances of innovative solutions*
> *Possibilities of breakthrough inventions*

REVOLUTIONARY ACCIDENTAL DISCOVERIES

Serendipity: ***accidental sagacity; the faculty of making fortunate discoveries of things you were not looking for***

DYNAMITE
Insulating nitroglycerin with volcanic soil for safe transport instead absorbs into and tames the deadly explosive

RAYON
Long thin strands of silky fiber form while wiping up a sticky spill

STAINLESS STEEL
Rust and chemical resistant steel born out of weathered alloy junk pile

MASONITE®
A leaky valve accidentally transforms waste wood into a sturdy and inexpensive hardboard product

OOPS!?

DRY CLEANING
Cleaning up spilled turpentine on a tablecloth brings a way to launder clothes without the use of water

SCOTCHGARD®
Spill on canvas shoes results in an oil and water repellent fabric protector

DYE INDUSTRY
Eighteen year old looking for synthetic quinine establishes world's dye industry

NITROGLYCERIN
Searching for a headache remedy, scientist was horrified to stumble upon such a deadly explosive

BAKELITE®
Inventor looking for synthetic shellac instead finds the first plastic—both valuable and inexpensive

MATCHES
Annoying glob on end of stirring stick fires up match industry

NOTES:

"We are continually faced by great opportunities brilliantly disguised as insoluble problems."

Lee Iacocca

THE CORNER OF HOLLYWOOD AND DESIGN

Designing new worlds in entertainment using product development tools, tactics, and talent

What does the perfect storm look like? Nobody knows since everyone who's experienced one has died in it. So Industrial Light + Magic faced an intriguing challenge when they developed the special effects for the film *The Perfect Storm*: create that which nobody has ever seen and yet make it look 'real.' After all, everyone *thinks* they know what it would look like. But how do you accomplish such a task? By bringing together a broad range of tools and experience. Obviously there will be animators, matte artists, and videographers involved, but what about an engineer to determine how a boat would respond to a huge wave? ILM brought engineers, computer programmers, industrial designers, and several other disciplines together to solve problems with this special-effects assignment. In today's world, entertainment is a product, and ILM uses industrial designers and engineers in much the same way that a manufacturer would. Filmmakers really do meet at the corner of Hollywood and Design.

TOOLS, TACTICS & TALENT

- > *Talent from design, engineering, science, and film*
- > *State-of-the-art hardware and software tools*
- > *Aggressive use of simulation and visualization*
- > *Experiences seen as being designed*
- > *Products seen as being experiences*
- > *Art, science, and engineering collaboration attitude*

RESULTS & BENEFITS

- > *Products that look and feel good, as well as work*
- > *Positive and exciting product/film experiences*
- > *Optimized balance of realism and fantasy*
- > *Products as entertainment for user*
- > *Improved market appeal and innovation content*
- > *Optimized virtual product development*

Industrial Light + Magic: San Rafael, California

Industrial Light + Magic is the arguably the world leader in film special effects and has revolutionized the way motion picture images are created. Located in Marin County and founded in 1975 by George Lucas, ILM has developed numerous innovative techniques for motion pictures. Originally established to provide technology for Star Wars, ILM has implemented computer graphics and digital imaging into its film making to create remarkable episodes. ILM has supplied visual effects for over 100 feature films and been involved in six out of the top ten box office hits of all time.

The Mask (New Line Cinema)

Snake Eyes (Paramount Studios)

Spawn (New Line Cinema)

Twister (Warner Brothers and Universal Pictures)

NOTES:

Art and Engineering SFX: Designing movies requires more than knowing film

Habib Zargarpour's hero is Leonardo da Vinci, and his resume emulates da Vinci's cross-disciplinary interests. Habib has degrees in both mechanical engineering and industrial design. He uses both degrees, plus skills in computer programming, sketching, and particulate physics, among others, to create the astounding special effects for which ILM has become famous. His recent work on the hit movie, *The Perfect Storm,* follows numerous others, including *Twister*, *The Mask*, *Spawn*, *Star Trek Generations*, *Star Trek First Contact*, *Jumanji*, and *Star Wars Phantom Menace*. Habib combines his design, mathematics, and engineering backgrounds in his work because they are absolutely necessary. "In the end," he states, "your eye is your only weapon." Particularly in a film like *The Perfect Storm*, you have to get the physics right, but that's not enough. It has to look right, too, or else the physics won't matter. This is also true in product design, leading to the need for more designers with backgrounds like Habib's.

Habib Zargarpour: Associate Visual Effects Supervisor, Industrial Light + Magic, San Rafael, California

Habib Zargarpour holds a degree in mechanical engineering from the University of British Columbia and a degree in industrial design from Art Center College of Design (graduating with distinction). He serves as associate visual effects supervisor for Industrial Light + Magic. Previous to his involvement at ILM, he owned his own visual effects company. He is a hands-on special effects specialist and has developed proprietary software for particulate modeling and other high-end physics applications.

Habib's dual film/engineering library

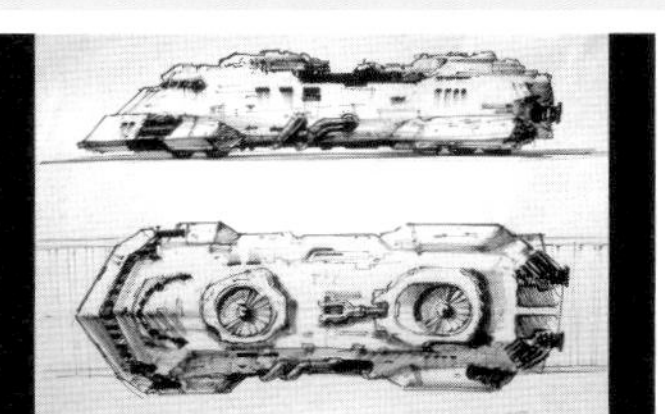

Advanced vehicle concept

Futuristic train concept

Rough and finished sea action models

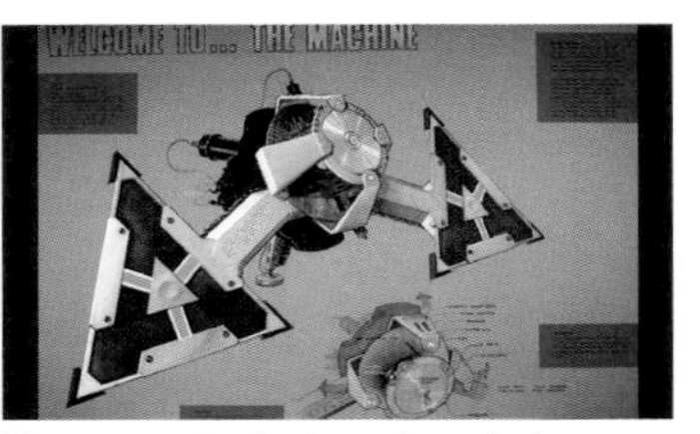

Electromechanical product design

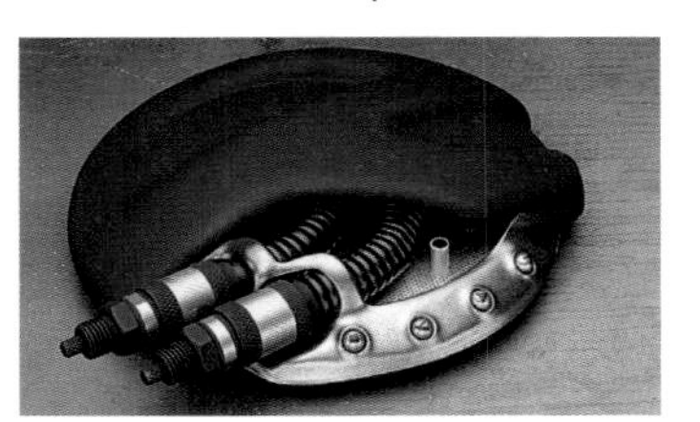

High-technology product design

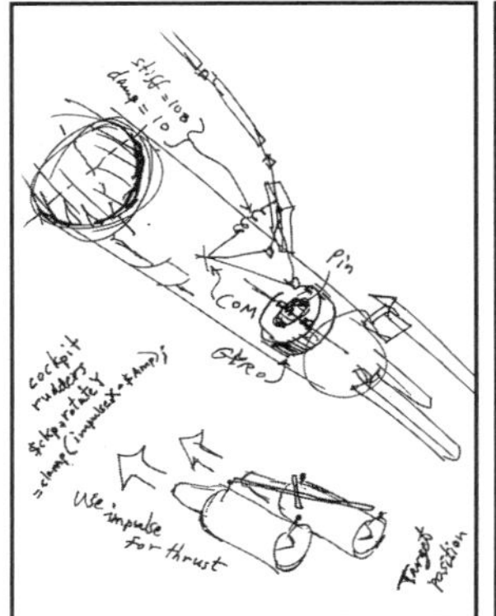

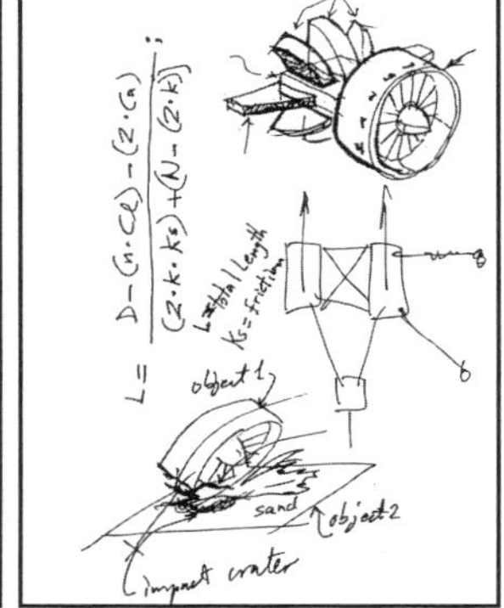

Pod racer sketches and physics formulas for film simulation

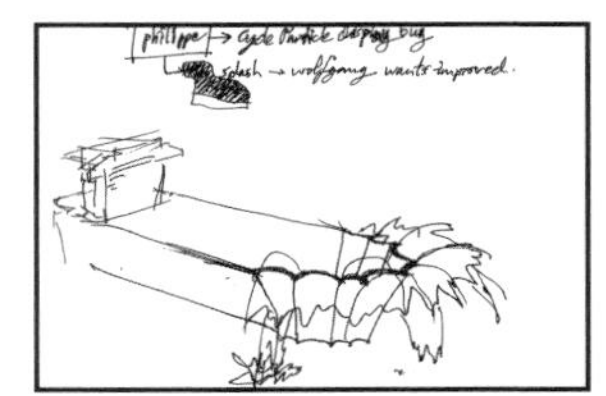

Sea wave motion sketches

Habib with his tools

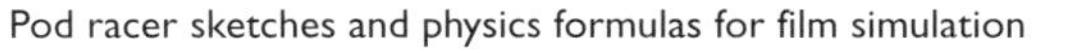

NOTES:

"*Chance favors only the minds that are prepared.*"
Louis Pasteur

ANALYSIS DRIVEN DESIGN

A process of sophisticated tools and tactics that helps optimize product design

Market pressures drive many industries to cut product development time. In general, consumer and even high-technology product developers have traditionally relied on testing physical prototypes later in the design process rather than on early computer simulations and virtual analysis. Unfortunately, the former frequently ends in locking the product to a less than optimum design. The product development process can be improved and compressed by starting simulated analysis early to develop guidelines that help get it right the first time and allow design changes at a time when they are less costly. Analysis can then continue as the design develops, guiding the process and design refinement. By the time physical prototypes are built, the product performance is well understood.—*Adapted by the author from SDRC material by Dr. Mary Baker*

TOOLS, TACTICS & TALENT

> *Integrated simulation, test, and analysis tools*
- *structural optimization and dynamics*
- *coupled loads analysis*
- *thermal, vibration, and acoustic analysis*
- *dynamic testing and evaluation*

> *Adequate computer workstations and accessories*
> *Capable computer test and analysis specialists*

RESULTS & BENEFITS

> *Optimized overall design*
> *Increased development speed*
> *Specifications met more precisely*
> *Reduced costs and resources required*
> *Improved product performance*
> *Reduced physical prototyping and testing*
> *Early problem detection and resolution*

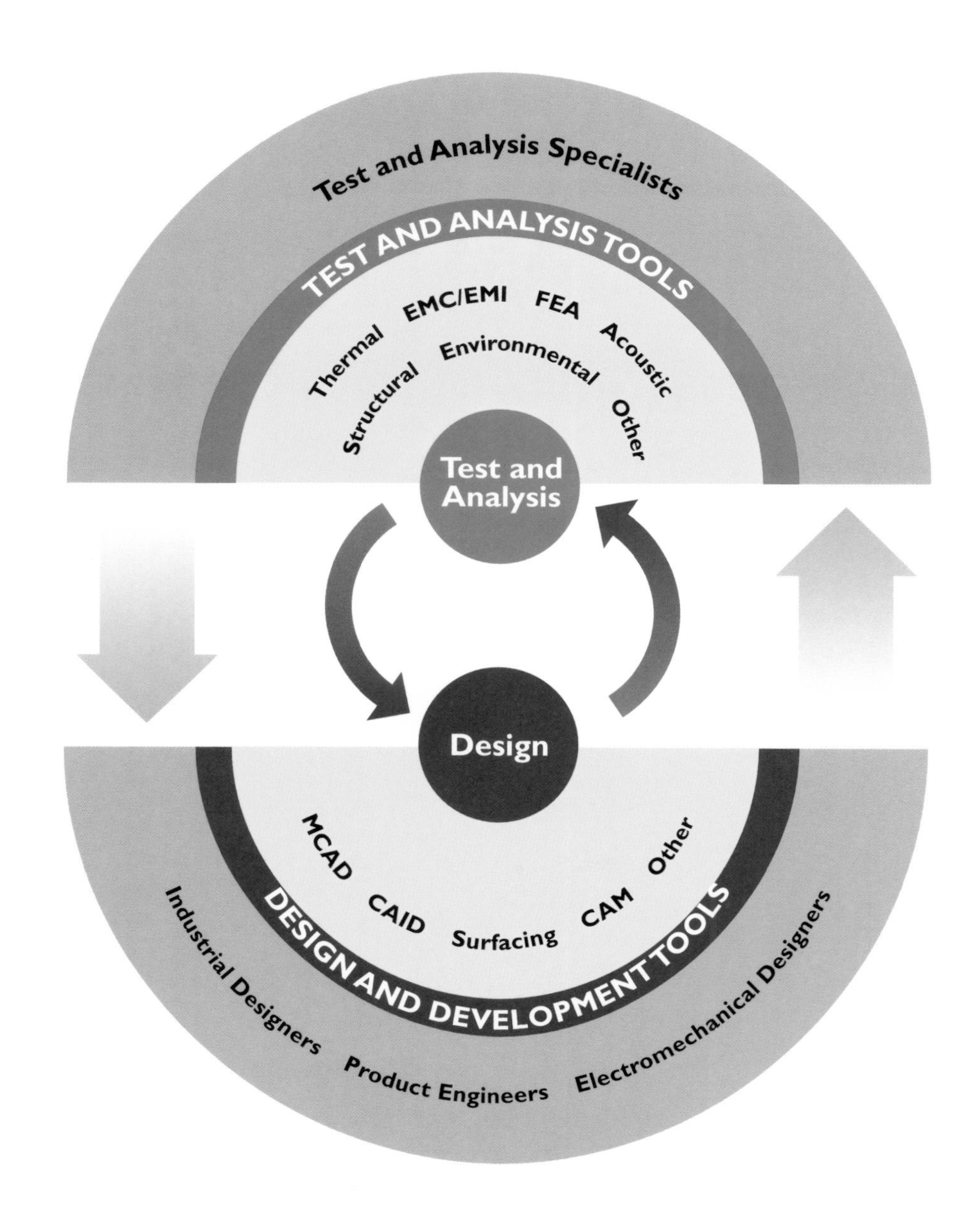

NOTES:

Thermal Design Optimization: An iterative and collaborative process for resolving difficult design problems

With ever-increasing system miniaturization where high-technology products are getting smaller with added features, capabilities and power, the issue of thermal management becomes increasingly important. Products such as small, compact personal data/video projectors have large thermal sources and heat sensitive components. In developing InFocus' revolutionary ultra light LP330 (code-named Dragonfly), a remarkably small projector, the thermal design challenge was enormous. Arc lamps in these units produce huge amounts of thermal energy. This heat must be dissipated while meeting stringent regulatory requirements for performance and safety. Gary Basey and Dr. Cathy Biber were part of a cross-functional project development team that devised a way to dissipate over 100 watts of heat while achieving very low fan noise levels. This was a engineer/analyst relationship with early and intimate collaboration using state-of-the art CAE and CFD analysis tools that resulted in a successful outcome.

TOOLS, TACTICS & TALENT

- > *Active use of Analysis Driven Design (ADD)*
- > *Highly interactive, collaborative team relationship*
- > *Powerful hardware and software analysis tools*
- > *Strong integration between software platforms*
- > *Cyclic and iterative design/test team attitude*
- > *Optimal use of physical and virtual modeling*

RESULTS & BENEFITS

- > *Early broad-based virtual design optimization*
- > *Optimized overall thermal design solution*
- > *Faster resolution of a critical development issue*
- > *Improved product environmental performance*
- > *Increased product safety for consumer/user*

InFocus Corporation: 27700B SW Parkway Avenue, Wilsonville, Oregon 97070-9215

Dr. Cathy Biber

Gary Basey

Cathy Biber, Ph.D., has more than ten years of experience applying analytical and experimental techniques to problem solving in product development and process design. She enjoys teaching seminars on applied thermal design in a wide range of industries to audiences all over the U.S. and Europe.

With companies such as Intel and Tektronix, Gary Basey brings over twenty years of product design and development experience to his position as a senior product design engineer at InFocus Corporation.

InFocus Corporation is the world leader in the design, development and manufacture of business and personal data/video projectors.

Physical SLA thermal test model

Optical engine subassembly to be thermally managed

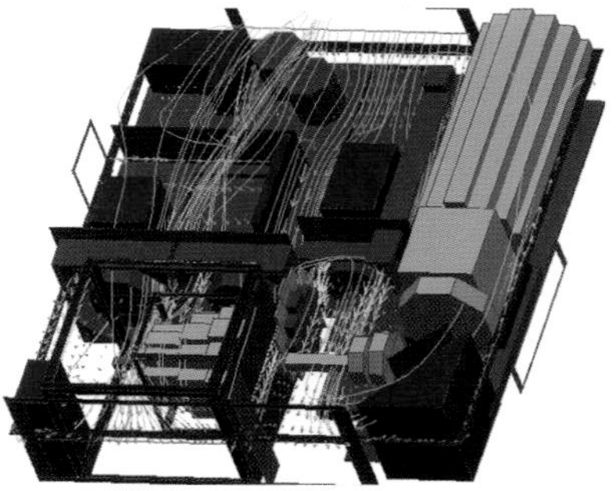

Virtual thermal CFD analysis model

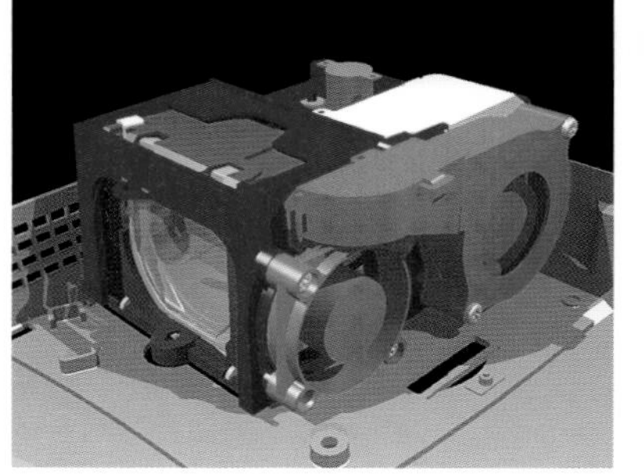

Virtual MCAD model of optical engine

Final Dragonfly production unit

NOTES:

"Tell me and I'll forget. Show me and I'll remember. Involve me and I'll understand."

Confucius

SPONSORED PROJECTS

Utilizing special programs at world-class colleges for corporate innovation

A highly effective tool for acquiring innovative ideas is to sponsor projects at quality design and development schools. The sponsored project approach not only taps into fresh ideas from talented students but provides unique and rewarding interaction for both the corporation and the educational institution. Sponsored projects may be implemented in every area across the enterprise, whether in design, business, operations, or manufacturing. There are a number of schools that already have such programs in place and have run smoothly for years. However, starting new programs at new schools or departments can be initiated with the proper support and interaction. World-class colleges such as Stanford University and Art Center College of Design have successfully worked with corporations in this venue for many years.

TOOLS, TACTICS & TALENT

> *Identify world-class design/development schools*
> *Appoint project leader and team internally*
> *Clarify objectives with comprehensive brief*
> *Be pro-active and involved throughout project*
> *Be certain core technology is understood*
> *Provide support resources and materials liberally*
> *Interact with faculty and students frequently*

RESULTS & BENEFITS

> *Team-building exercise for both sides*
> *Contribution to development education*
> *Inexpensive innovation with high ROI*
> *Abundant, fresh, new ideas and approaches*
> *Potential design talent pool established*
> *Exposure to new tools and tactics on both sides*

Art Center College of Design: 1700 Lida Street, Pasadena, California 91103

Art Center College of Design is internationally recognized for its programs in art and design. Located in Pasadena, California, ACCD is a world leader in industrial design, digital design, and media innovation education. It was founded in 1930 in Los Angeles and offers B.S., B.F.A., and M.F.A. degrees in a full spectrum of visual arts and design disciplines. ACCD is dedicated to professionalism and its faculty members help students to develop outstanding conceptual ability and technical awareness. Many ACCD graduates are design leaders and executives in prominent design organizations throughout the world.

State-of-the-art campus

Interactive environment

Talented student body

Students and wall of sketches

Full-scale design projects

Concept model and mockup shops

Vehicle design and modeling labs

Crafted, detailed execution (H. Takahashi)

NOTES:

Fresh Creative Ideas: Sponsored projects can be applied to a broad spectrum of development

Art Center College of Design in Pasadena, California has had a sponsored project program for many years. Many of its sponsored projects are in the areas of product, transportation, environmental, and entertainment design. Both small and large corporations engage the design students in challenging and rewarding experiences. Such companies include Honda, General Motors, Ford, InFocus, Sony, Nokia, JPL, and a multitude of others. These projects are primarily of a conceptual and future-oriented nature rather than for resolving specific current design problems. This philosophy forces both corporate and student teams to think out of the box and into the future. Represented on this page are a number of examples of sponsored projects from this excellent school.

"Art Center's educational philosophy has always been to prepare creative individuals for the professional world. In order to do this, it is essential that we push the boundaries of a design and art education, to fully explore the creative process that will result in the most creative solutions of the future. The sponsored project program provides students with an opportunity to develop solutions to specific design issues put forward by each sponsor. The program functions as a catalyst for new, energetic ideas exchanged between the classroom and the outside world resulting in mutual benefits to not only our graduates, but to their eventual clients, employers, and ultimately to the world at large."—Richard Koshalek, President, Art Center College of Design

Richard Koshalek: President, Art Center College of Design (ACCD), Pasadena, California

Richard Koshalek began his tenure as the fourth president of ACCD in September, 1999. He attended the University of Wisconsin, obtaining a B.A. in architecture, and the University of Minnesota, receiving an M.A. in architecture and art history. Before joining ACCD, Mr. Koshalek served as chief curator, deputy director, and director at several prominent museums in the U.S. He was most recently the director of the Museum of Contemporary Art in Los Angeles. Mr. Koshalek is a member of the American Association of Art Museum Directors, the Chase Manhattan Bank Art Committee, and a board member of the American Institute of Graphic Arts.

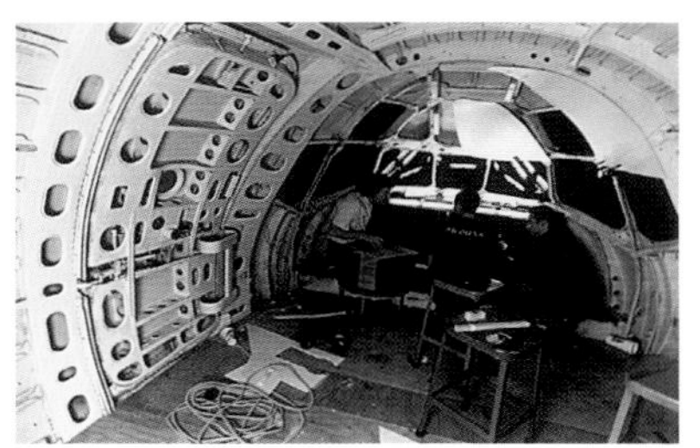

Lockheed cockpit installed at ACCD

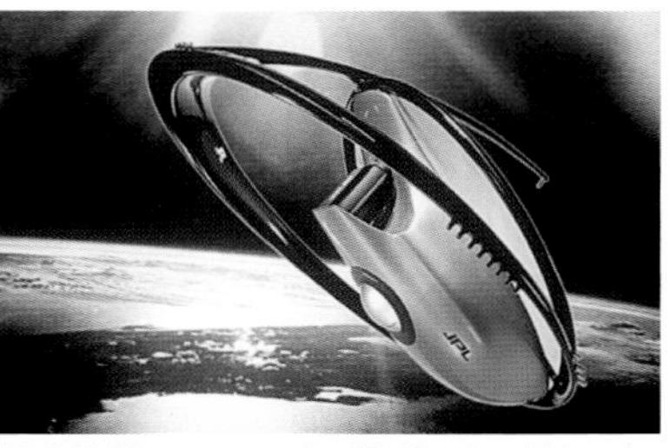

JPL/NASA space probe (C. Kember, T. Levy)

Northrup medical module (J. Meyer)

Peugeot advanced vehicle (D. Nagao)

Aprilia motorcycle design (A. Markevich)

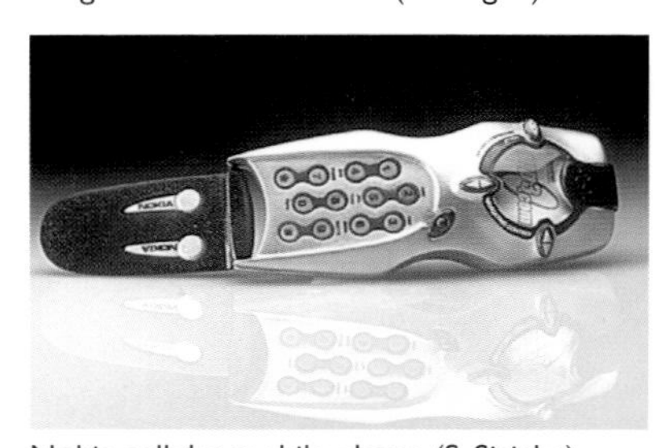

Nokia cellular mobile phone (S. Siricha)

Children's alarm clock/radio (R. Garcia)

Reebok footwear design (D. Dombrow)

NOTES:

VIRTUAL INTEGRATION

"The essence of competitiveness is liberated when we make people believe that what they think and do is important—and then get out of their way while they do it."

Jack Welch, CEO of General Electric

A development environment where design intent is intact from concept to production

Industrial design and product engineering have traditionally used different product development tool sets. Communication between them has historically been challenging. This dichotomy is true in other areas such as design engineering and manufacturing engineering, electronics engineering and mechanical engineering, and other discipline pairs that must communicate and integrate their designs successfully across functional boundaries. Even with the advent of computer-aided engineering and design in virtually all disciplines, there has still been difficulty in communication and integration until recently. Today it is possible to have a variety of functional groups effectively transferring design data between platforms. In a virtual environment the appearance surfaces of the industrial designer can be communicated and translated effectively to the mechanical designer with intact design intent, and the mechanical designer's completed parts and assemblies can be transferred to manufacturing for tool fabrication and mass production.

TOOLS, TACTICS & TALENT

> *Compatible design and development tools*
> *CAID, surface, MCAD, and CAM software*
> *Staff attitude and training for integration*
> *State-of-the-art support resources*
> *Mutual functional appreciation and support*
> *Highly talented and compatible design specialists*

RESULTS & BENEFITS

> *Smooth, seamless data transfer*
> *Optimized hand-offs between disciplines*
> *Minimized rebuilding of surfaces and models*
> *Retention of original design intent*
> *Minimized conflict between functional groups*
> *Optimized project schedules*

Thomas Heermann: Industry Marketing Manager, Alias|Wavefront, Toronto, Canada

Thomas is the industry marketing manager for the Design Division at Alias|Wavefront. Prior to joining A|W, he was responsible for marketing and communications at ICEM Technologies, a developer of Class-A surfacing software. Seamless integration of CAID, surfacing, MCAD, and CAM development tools for the disciplines of industrial design, mechanical engineering, and manufacturing is of special interest to Thomas and occupies an important portion of his work. He holds a degree in engineering and business from FHT Esslingen. Born in Stuttgart, Germany, Thomas now lives in Toronto.

CROSS-FUNCTIONAL VIRTUAL INTEGRATION DIAGRAM

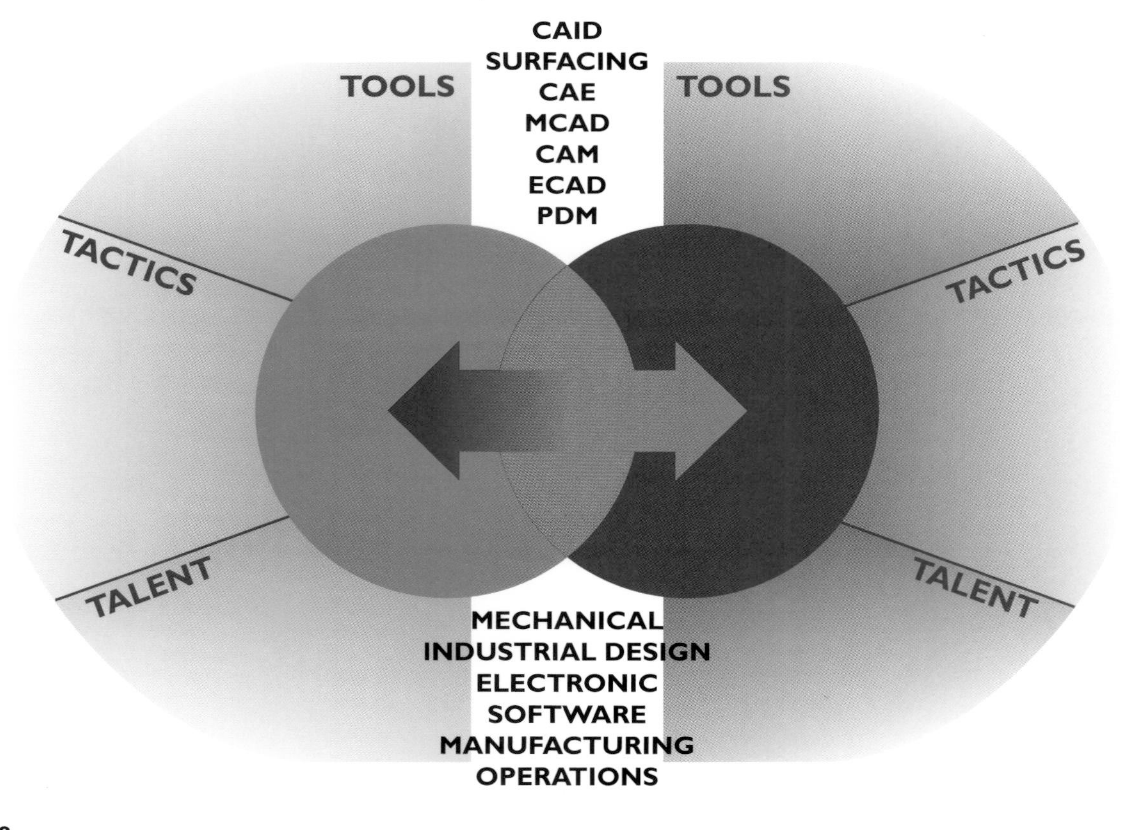

NOTES:

Surface Integration Success: Industrial design and mechanical engineering integrating across platforms

During the design of a sophisticated high-technology product having complex external surface forms and features developed by industrial design, these surfaces need to be transferred to product engineering for integration into the virtual mechanical design. An unfortunately common process for doing this is where engineering, after receiving the surfaces from industrial design software, only uses these as patterns for completely rebuilding the surface in the MCAD modeler. Designers and engineers at Lexmark have streamlined this process in developing their printer designs. The industrial design surfaces are directly transferred to the engineering MCAD modeler and used intact without rebuilding the surfaces. In this way, not only is the original industrial design intent captured and retained but the process is very efficient since engineering need not redevelop and rebuild a surface already completed.

John Gassett: Senior Industrial Designer, Lexmark International, Inc., Lexington, Kentucky

John has been with Lexmark since it was established in 1992. He was a key contributor to Lexmark's design language for network laser printers and has designed a multitude of products including printers, network adapters, furniture, and IR devices. Before working for Lexmark, John designed printers for IBM's Printer Division. John has a B.S. in industrial design and is a member of IDSA. He holds a number of design patents and his design for the Optra S is in The Chicago Athenaeum Museum of Architecture and Design's permanent collection.

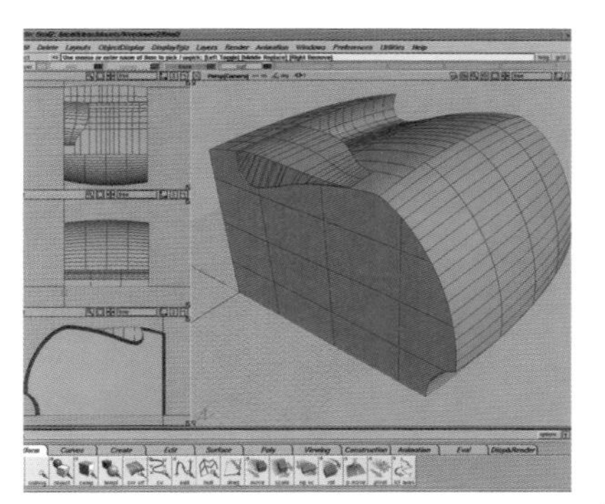

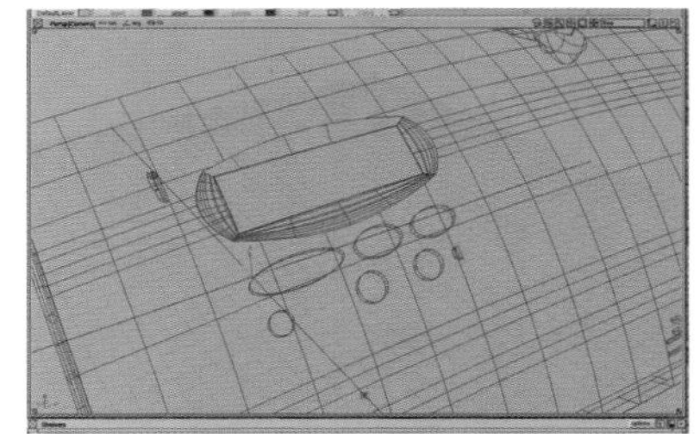

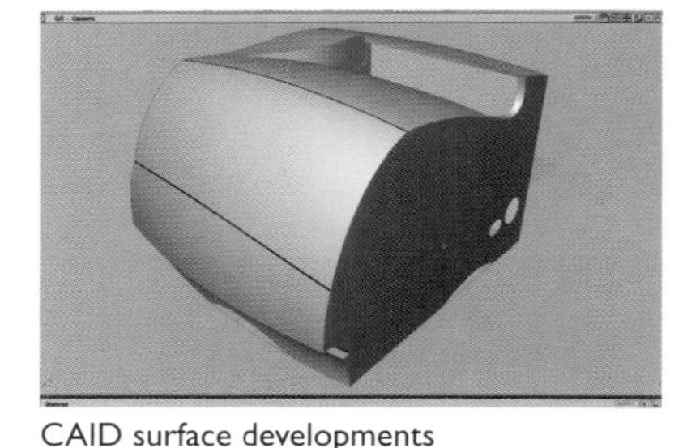

CAID surface developments

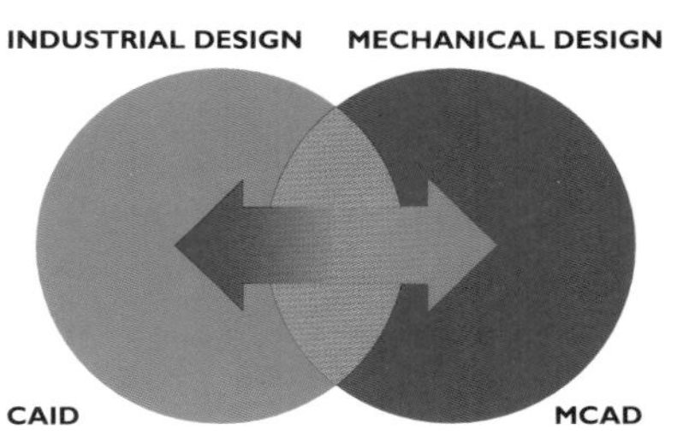

Integration process

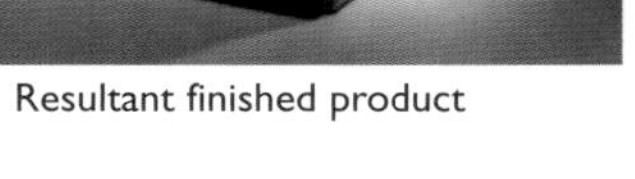

Resultant finished product

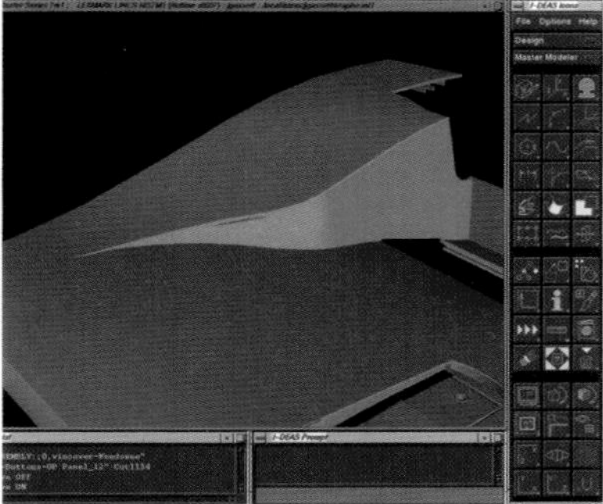

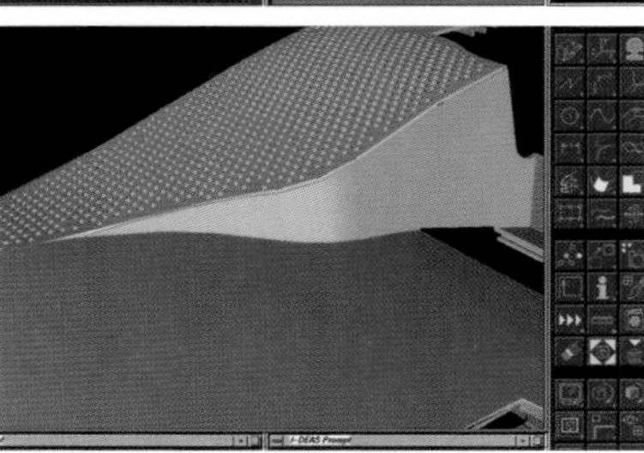

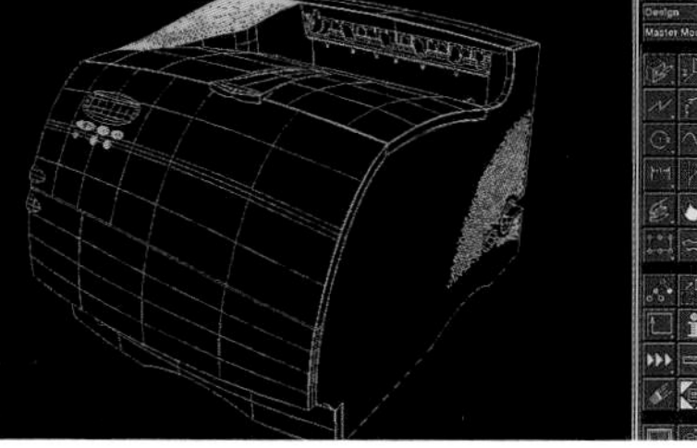

CAE/MCAD solid models

NOTES:

Concurrent Integration: Simultaneous internal and external virtual product design and development

An integration issue that very frequently emerges in "concurrent" product development is the interface and data transfer between industrial design and product engineering. Often the industrial designer is using a CAID surfacing software tool to develop the product exterior features and cosmetics and the product engineer is using an MCAD software tool to develop the internal mechanical design and assembly. Alchemy is a design consulting firm that has streamlined this interface and the concurrent interaction and integration of design modeling data between the two functions. In developing the Vortex toy gun, Alchemy seamlessly integrated the internal mechanical structure and mechanism design MCAD models with its own exterior surface form and ergonomics CAID models literally days before shipping the integrated model data for tooling.

TOOLS, TACTICS & TALENT

- > *Close understanding and working arrangement between industrial design and product engineering*
- > *Compatible state-of-the-art design tools and communication processes between functions*
- > *Well-defined system of design integration*
- > *Mutual appreciation of different design functions*
- > *Manufacturing suppliers with state-of-the-art compatible software tools*
- > *Highly competent and skilled design talent*

RESULTS & BENEFITS

- > *Optimized development schedule time line*
- > *Design creativity/innovation both outside and inside*
- > *Fully integrated cross-functional design solutions*
- > *Minimal design integration problems*
- > *Optimized manufacturing hand-off and resolution*

Alchemy: San Francisco, California

Alchemy was founded by Gray Holland and has recently been merged with frogdesign of San Francisco, California. Gray is now a vice president at frogdesign. Alchemy's forte was in digital industrial design and the use of sophisticated virtual design tools to develop world-class products. Alchemy created designs and visualizations for a variety of products, including concepts, models, and animations for development, marketing, and branding purposes. The Vortex toy gun was co-designed with Function Engineering of Palo Alto, California.

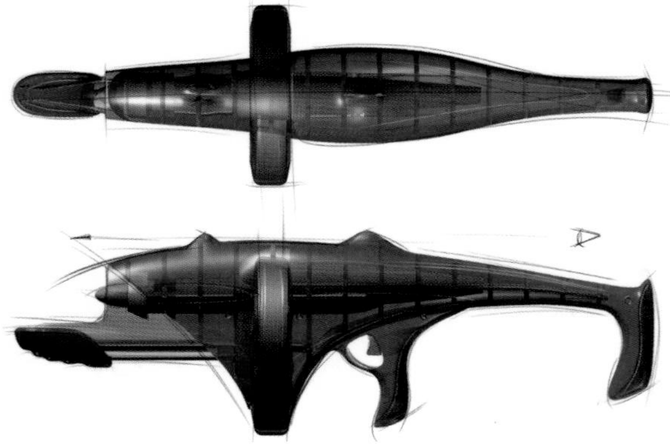

Refined industrial design exterior and features in CAID

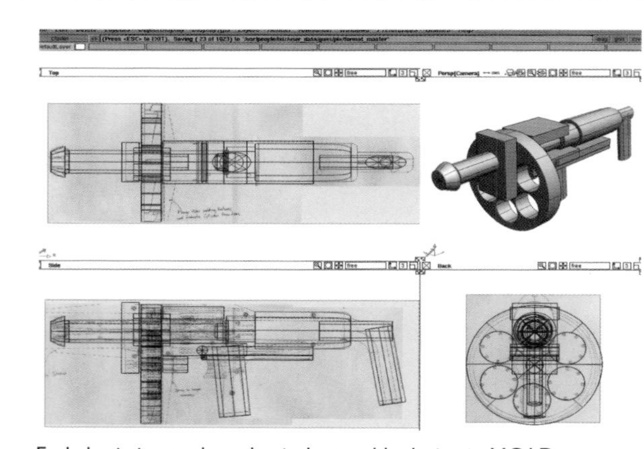

Early basic internal mechanical assembly design in MCAD

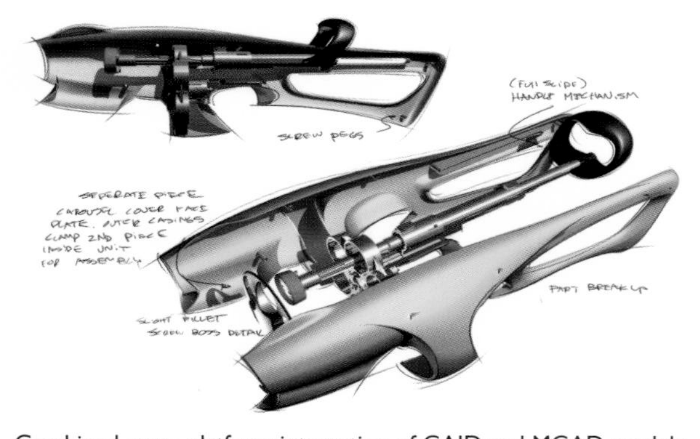

Combined cross-platform integration of CAID and MCAD models

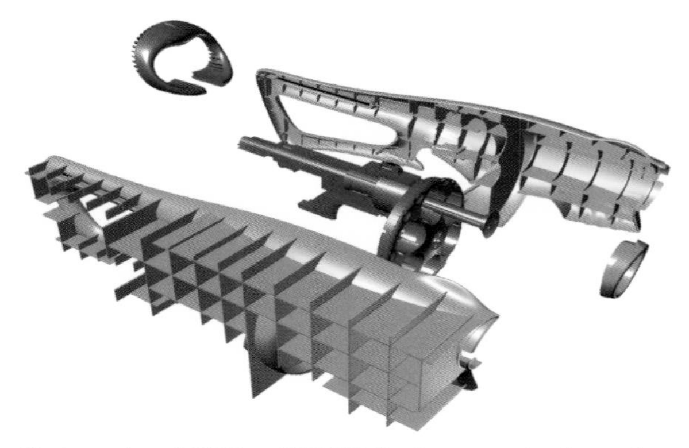

Final merging of CAID and MCAD files in preparation for tooling

NOTES:

Multiple Media Integration: Using a full spectrum of virtual development tools to design complex systems

The optimum development integration process would be one where all forms of virtual simulation are utilized seamlessly and appropriately together. The Ethicon industrial design group approaches this challenge quite well. They utilize a broad spectrum of visualization and simulation tools and techniques. These include freehand sketching and illustration, digital sketching, industrial design modeling and rendering, MCAD modeling, and compositions and animations utilizing combinations of these. This process is part of the development of their medical/surgical products, their applications, and training in procedures. The virtual results include overall products and details, operating room procedures, full product assemblies, and animations of applications in surgical and medical procedures. These are also utilized successfully in promotion, advertising, and marketing applications.

TOOLS, TACTICS & TALENT

> *Variety of designer capabilities and expertise*
> *Utilization of multiple image/design media*
> *State-of-the-art simulation/integration processes*
> *Visualization specialists for presentation*
> *Commitment to high visual content in all areas*
> *Best-in-class hardware and software tools*

RESULTS & BENEFITS

> *Increased innovation and creativity of designs*
> *More comprehensive development process*
> *Higher communication and understanding*
> *Improved user/customer interaction and response*
> *Better management visibility of process/results*
> *Improved marketing and sales presentation*

Ethicon Endo-Surgery, Inc: 4545 Creek Road, Cincinnati, Ohio 45242-2839

Ethicon is a Johnson & Johnson Company and a world leader in the manufacture of surgical instruments that specialize in minimally invasive surgery, an operational procedure that avoids making large, open incisions in the body. Ethicon's close relationship with prominent surgeons has helped to enhance the design and development of its products. Its well-known educational establishment, the Endo-Surgery Institute, has educated thousands of medical professionals in the latest surgical technologies. Ethicon prides itself in working with surgeons worldwide while designing beneficial new products and improving operating techniques.

Freehand sketch of surgical procedure

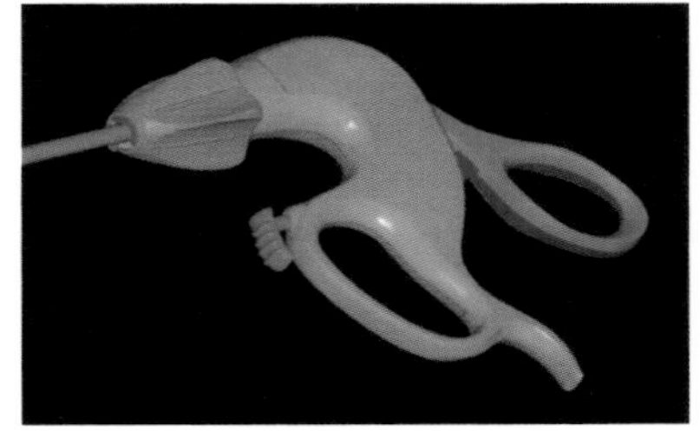

MCAD solid model of surgical device

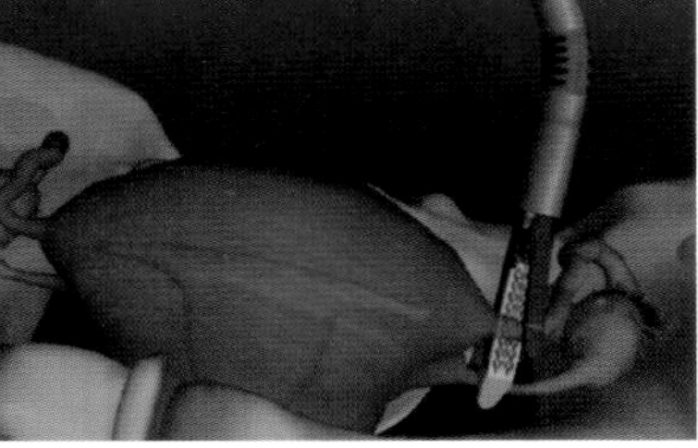

Animation of surgical procedure

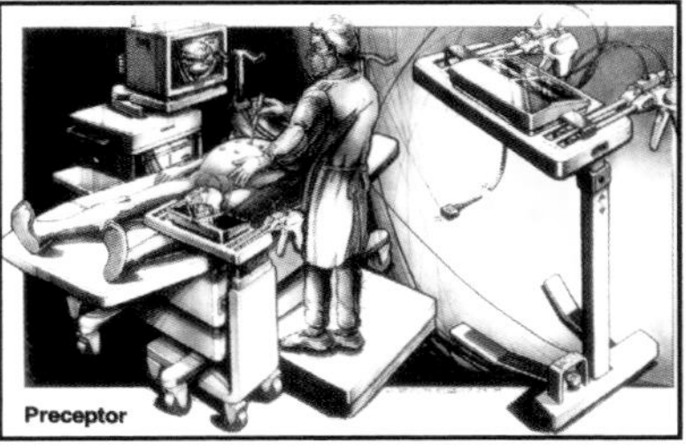

Freehand sketch of operating procedure

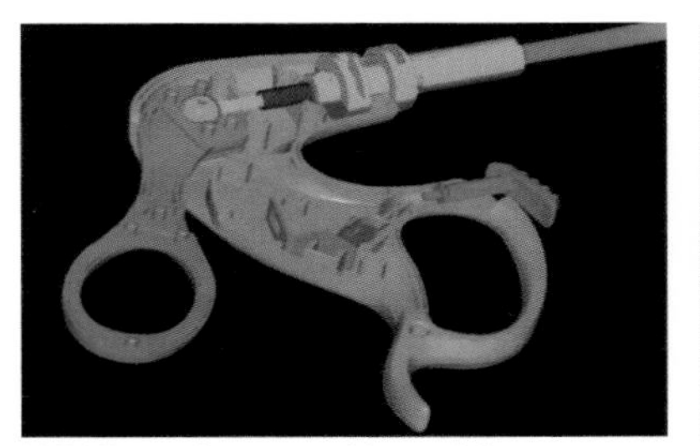

MCAD product assembly model

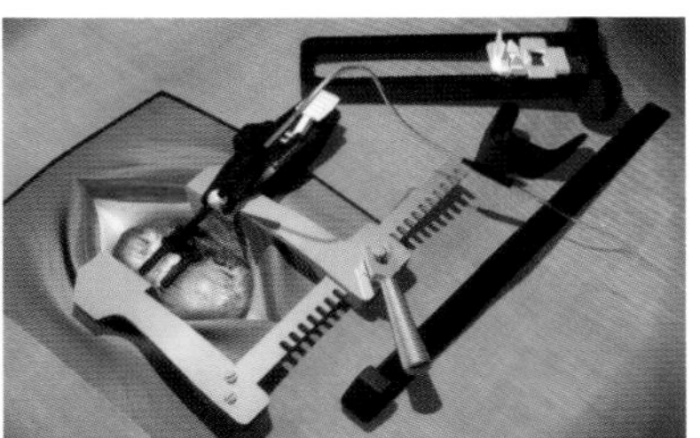

Animation of surgical device use

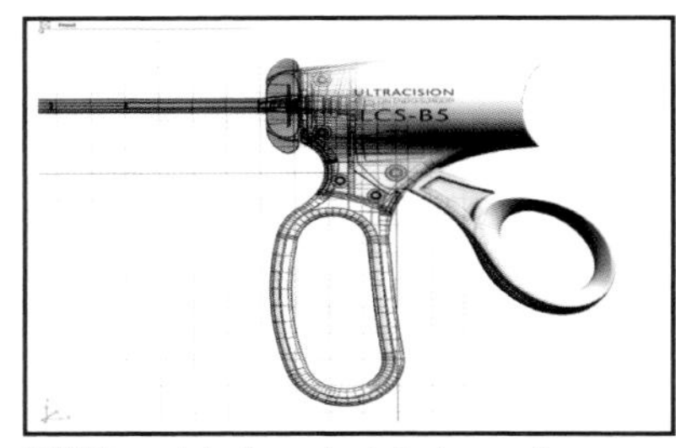

CAID model of product design surfaces

CAID product image for use in marketing

CAID/MCAD image of design structure

NOTES:

PRODUCT ARCHITECTURE

"In order to have actionable meaning, the fuzzy mental models in top management minds must ultimately be externalized in representations the enterprise can grasp."

Michael Schrage

The importance of early product configuration and physical system organization

The architectural or component configuration of a product can drastically affect its design success. Architectural configuration impacts many critical factors in a product's development. These include size, weight, compactness, efficiency, simplicity/complexity, manufacturability, cost, bill of material, agency compliance, and many other important issues. Too often, architectural configuration design (sometimes called, or included in, product system design) is ignored, neglected, or inadequately addressed until late in a project. This delay greatly impacts schedule and cost if late configuration changes are needed. Identifying all viable components and configuration options as early as possible will only benefit the overall product development.

TOOLS, TACTICS & TALENT

> *Ideation of multiple component configurations*
> *Physical and/or virtual modeling of all options*
> *Systematic configuration design process*
> *Interactive group evaluation process*
> *Configuration options capture methods*
> *Best modeling media for effective simulation*

RESULTS & BENEFITS

> *Optimized product architecture and configuration*
> *Optimal satisfaction of product requirements*
> *Minimized problem emergence later*
> *Early elimination of problematic solutions*
> *Increased innovation and new idea generation*
> *Best product size, weight, and features design*
> *Improved product performance and profitability*

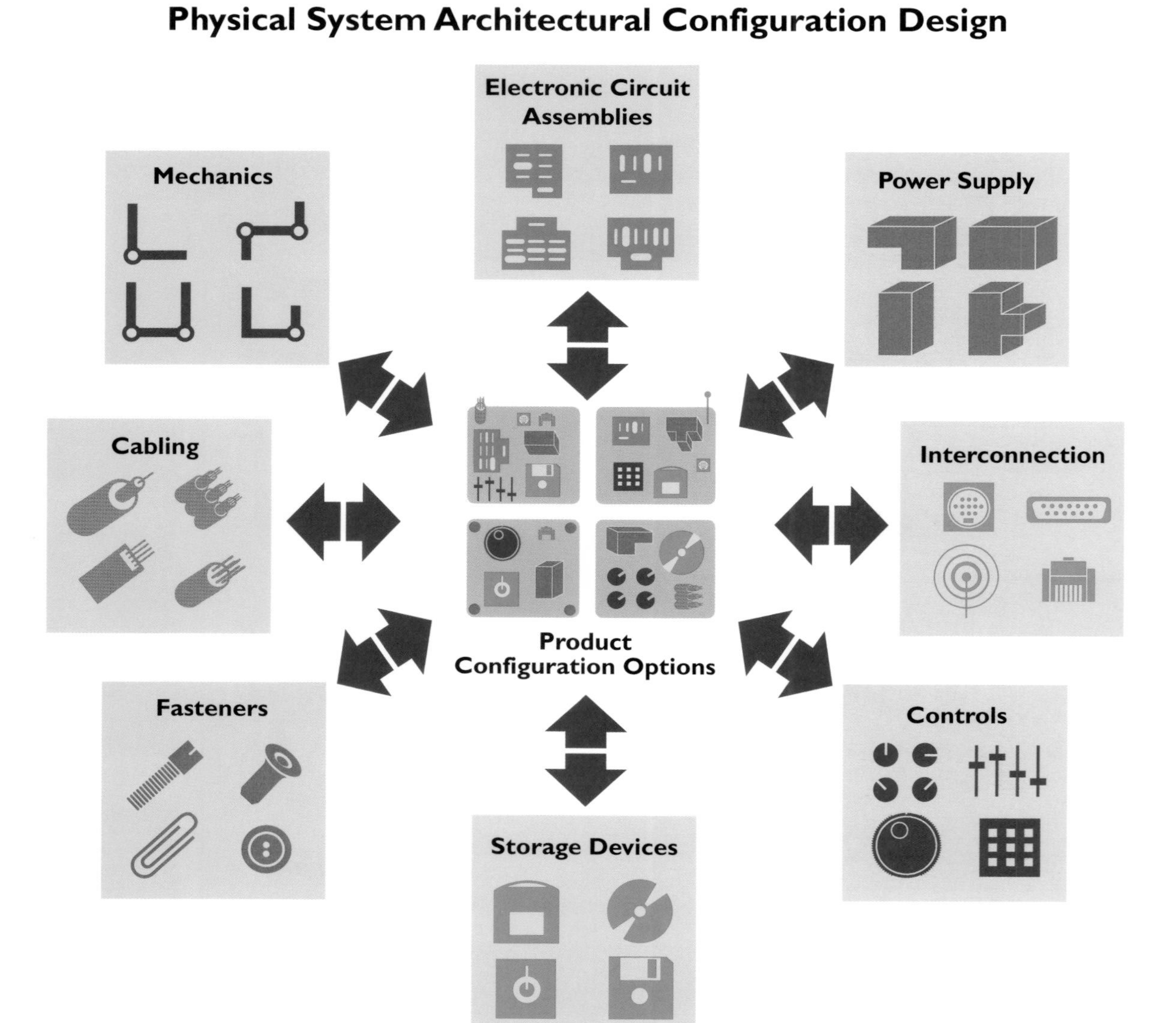

NOTES:

Product Configuration Design: A critical development process that must be iterated early in a project

During the early product architecture and component configuration process, there must be a system of tools, tactics, and talent that supports this effort effectively. Materials and processes that include simple paper doll layouts, foam blocks simulating components, actual hardware items and assemblies, and virtual computer configuration models are essential to optimizing configuration design and product architecture. Interaction with the components and their arrangements must be a combination of visual, tactile, and kinesthetic experiences utilizing both physical and virtual manifestations of the product system. Lexmark engineering uses such a variety of tools and tactics to configure their complex printing mechanisms, engineering assemblies, and complete systems. For whatever industry, whether it's an inkjet or laser printer, a data/video projector, a laptop computer, or any other compact and dense high-technology product system, only optimized architectural configuration design will suffice.

Larry Stahlman: Mechanical Development Manager, Lexmark International, Inc., Lexington, Kentucky

Larry is a mechanical development engineering manager for Lexmark International and was part of the transition from IBM to Lexmark in 1991. Larry began working for IBM in 1974 and held several management positions in product engineering, purchasing, human factors, and business plans and controls. He has managed numerous design teams on projects ranging from typewriters and keyboards to laptop computers and business printers. His current responsibilities at Lexmark cover a broad spectrum of development functions. Larry has a B.S. in mechanical engineering, has done graduate work, and holds several international design patents.

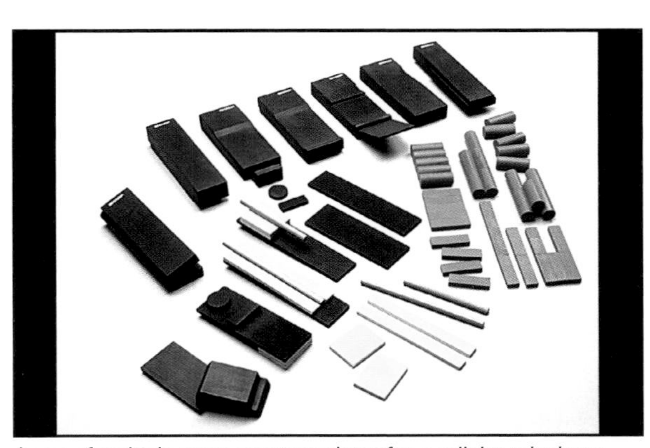

Array of multiple component mockups for a cellular telephone

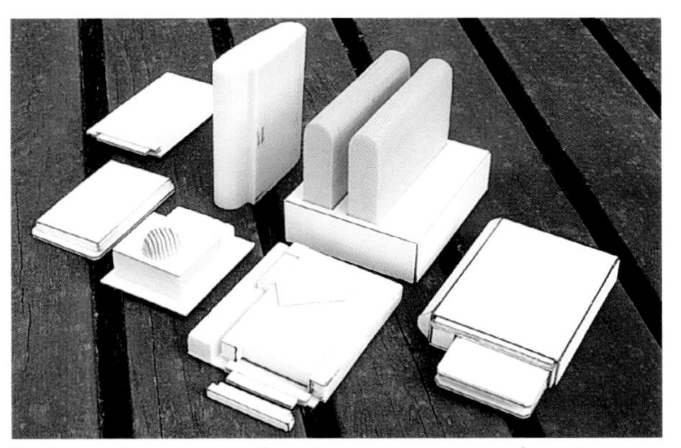

Mockups of components for a laptop computer configuration

Snapshot of one configuration option for an LCD projector

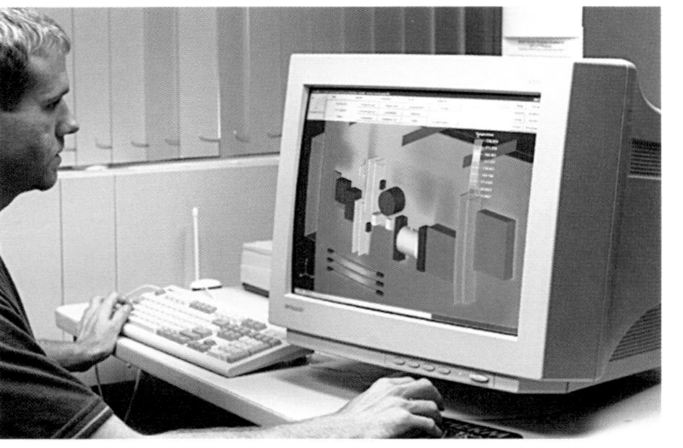

Manipulation of components using virtual 3D configuration

NOTES:

"We have to understand that the world can only be grasped by action, not by contemplation. The hand is more important than the eye . . . the hand is the cutting edge of the mind."

Jacob Bronowski

ONE-ON-ONE DESIGN

The combination of expert user and expert designer for new product innovation

Traditionally, the product development process has been rife with disconnects. Designers fought with engineers, who argued with production people, etc. But too frequently, the biggest disconnect has been between product development people and the end users of their products. These disconnects can no longer be tolerated, particularly the one involving the customer. Toward that end, Nike has been developing very close relationships with the athletes it sponsors, leading to products that are designed with the end user in mind. This is an innovative variation on the traditional product development cycle where an expert user helps shape the design of an initial product through close interaction with an expert designer. The expert user then puts the product to use in real life, identifying the product s strengths and weaknesses and developing a wish list for the next design iteration. The cycle continually refines the product, not on the whims of design trends but on the requirements of the expert user.

TOOLS, TACTICS & TALENT

> *Expert designer with strong skills*
> *Expert user willing to collaborate*
> *Chemistry between expert user and designer*
> *Drive design by needs of the expert user*
> *Define/refine design through expert designer*
> *Finalize design with final customer in mind*

RESULTS & BENEFITS

> *Product better suited to needs of marketplace*
> *Ongoing development cycle refines product*
> *Maintain market position by staying ahead*
> *Maximized value of sponsorships/endorsements*
> *Reduced time-to-market from intense expertise*
> *Translated needs of user into viable product*

Nike, Inc.: One Bowerman Drive, Beaverton, Oregon 97005-6453 (and worldwide)

Nike is an international market leader in athletic footwear, sportswear, and accessories and has been endorsed by a multitude of prominent athletes. Originally named "Blue Ribbon Sports," Nike was founded by Phil Knight, a former business major and track runner at the University of Oregon, and Bill Bowerman, the legendary track coach at the University of Oregon. Nike has an innovative development process of designing and marketing athletic shoes that requires an extensive collaboration of engineers, designers, developers, marketing experts, and athletes. Nike prides itself in its dedication to producing the highest quality product possible that not only helps the athlete perform at a higher level than their opponent but is also aesthetically pleasing.

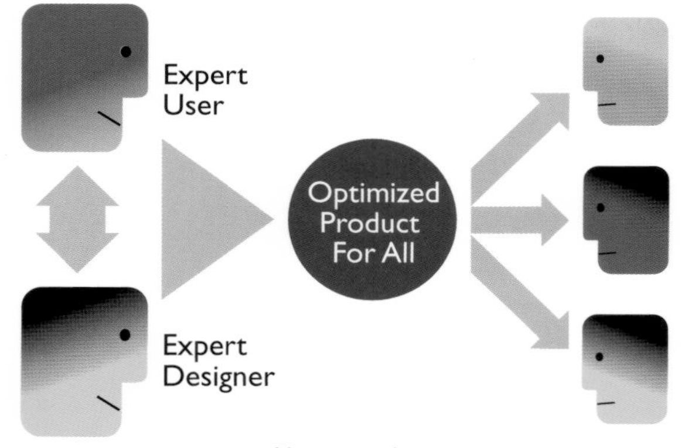

How it works

Aaron Cooper and Scottie Pippen interacting in hands-on product design process

NOTES:

Collaborative Expertise: The ultimate in design interaction

When an expert user of basketball shoes meets an expert designer charged with developing new basketball shoes, great things can happen. Such a relationship developed between Aaron Cooper, Nike industrial designer, and Scottie Pippen, six-time world champion power forward for the Portland Trailblazers. The first version of Pippen's signature shoe was based on the new relationship between designer and expert user. It was a very good shoe, but it was going to become better. As the relationship developed, Aaron came to better understand Scottie's approach to the game and the shoe. The more he understood about Scottie, the better the shoe met his needs. Now in its fifth iteration, the Pippen shoe has been recognized for its purity and simplicity of design.

Aaron Cooper: Senior Industrial Designer, Nike, Inc., Footwear Division, Beaverton, Oregon

What do Scottie Pippen, the NBA professional basketball player, and Aaron Cooper, industrial designer, have in common? More than you may think. Aaron is a senior designer at Nike Footwear and has personally designed all of Scottie Pippen's signature basketball shoes. Recently, Aaron's shoe design has been recognized by several magazines worldwide for its excellence. Aaron received his B.S. in industrial design from Art Center College of Design in Pasadena, California. When he is not designing sports shoes he enjoys spending time with his son, whom he considers to be his greatest portfolio piece.

Inspiration imagery

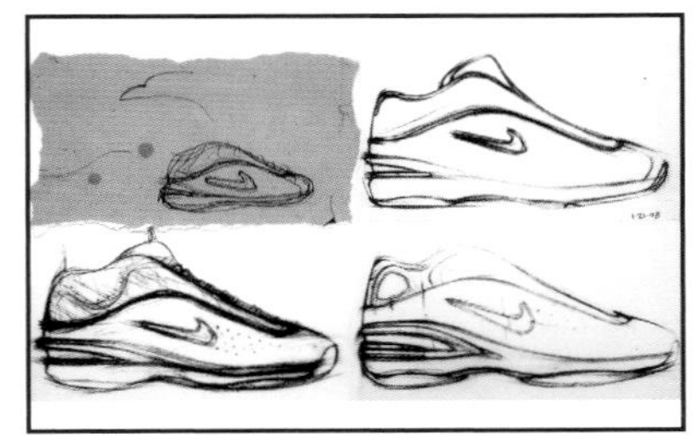

Concept sketches

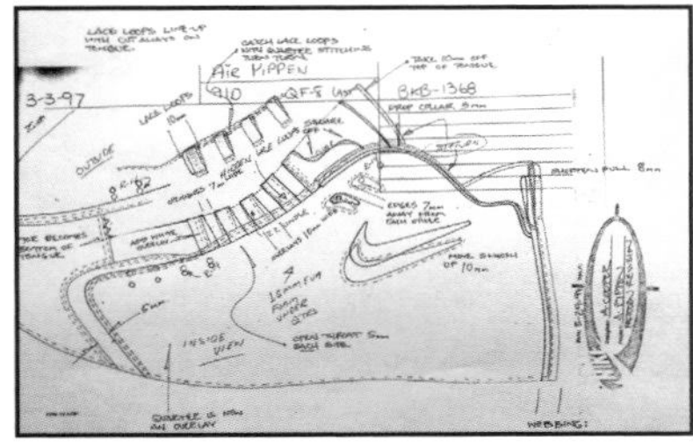

Detailed shoe layout

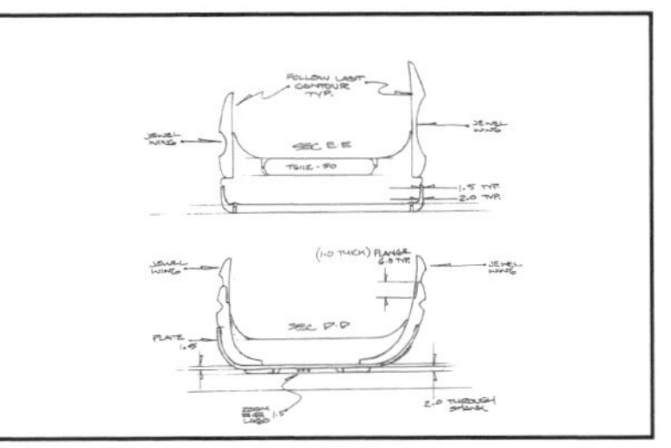

Cross-sectional view

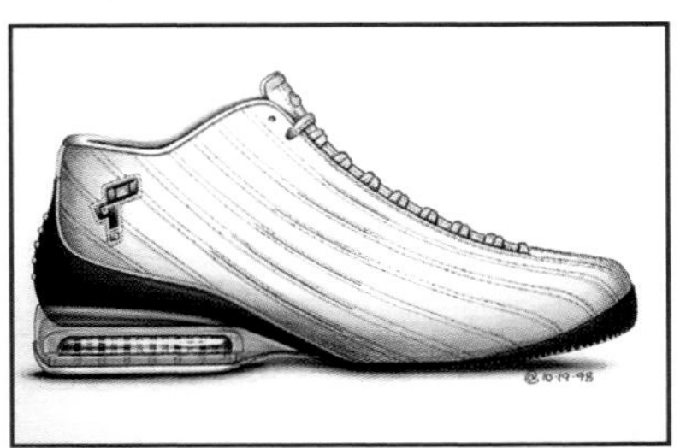

Final rendering of design

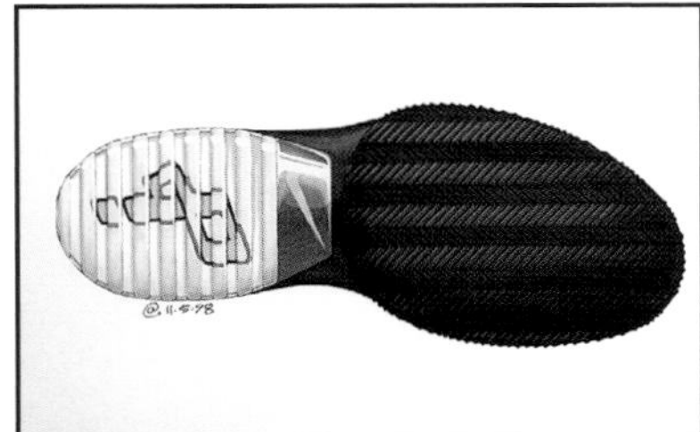

Sole design

Various shoe parts

A design evolution/revolution

The Shoe

NOTES:

"Labor is prior to, and independent of, capital. Capital is only the fruit of labor, and could never have existed if labor had not first existed. Labor is the superior of capital, and deserves much higher consideration."

Abraham Lincoln

"Does your company have a clean-desk policy? If so, the company's nuts and you're nuts to stay there."

Tom Peters in *The Tom Peters Seminar*

MANUFACTURING & OPERATIONS

"If you think the problem is bad now, just wait until we've solved it."
Arthur Kasspe

INFORMATION OVERLOAD

Multiplying enterprise information is in need of total management, control, and dissemination

Today's product development enterprise is inundated with mountains of data produced every day. This includes sophisticated CAE files from product development, corporate business and finance numbers, and operations and manufacturing information. Not only does there seem to be too much to manage, it is also distributed among a number of internal and external organizations. Worst of all, this information is often inaccessible to those who really need it except by painful extraction. A few bridges exist of varying types between departments or suppliers. However, they are often inefficient due to the different departmental software tools used to manage the data. In most cases, these tools are incompatible with one another and "converse" with difficulty, if at all. The result is that nearly anyone in the corporation who has a job to get done cannot optimize their performance.

PROBLEMS

- > *Multiple incompatible departmental data management tools*
- > *No one champion of a total access information system*
- > *Still using some form of paper documentation for drawings and records*
- > *Lack of comprehensive single database system for both internal and external organizations*

RESULTS

- > *Inefficient to disastrous information sharing*
- > *Inadequate information access to get work done*
- > *Everything slows down including revenue*
- > *Frustration and dissatisfaction of workers*
- > *Longer projects and compromised time-to-market*
- > *Reduced productivity and profitability*

Islands of Information and Automation:

discreet corporate information centers adrift in a sea of data chaos

Many creators and users in a sea of chaos . . .

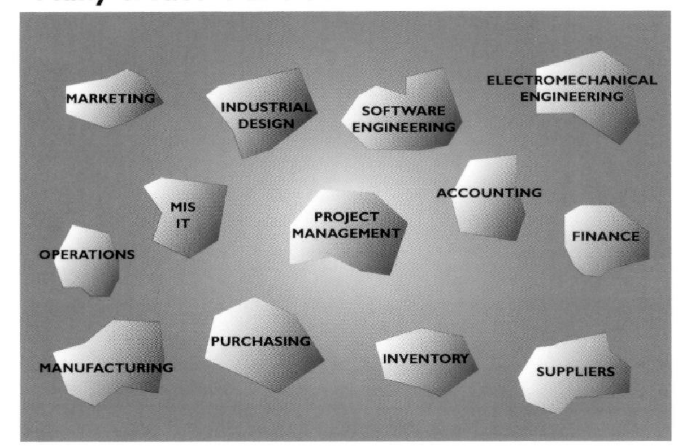

with a few bridges here and there . . .

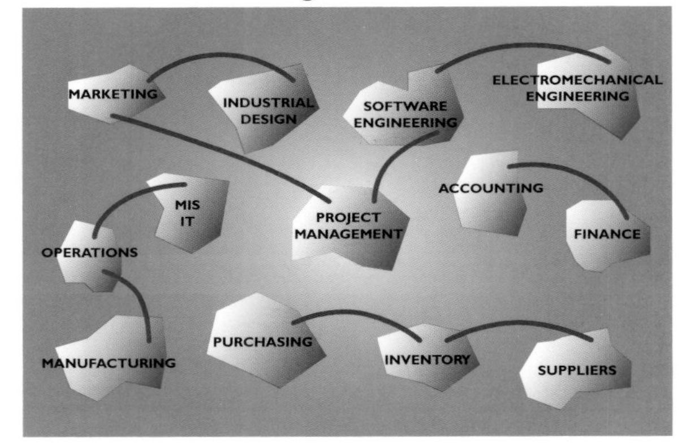

and a variety of "proprietary" tools . . .

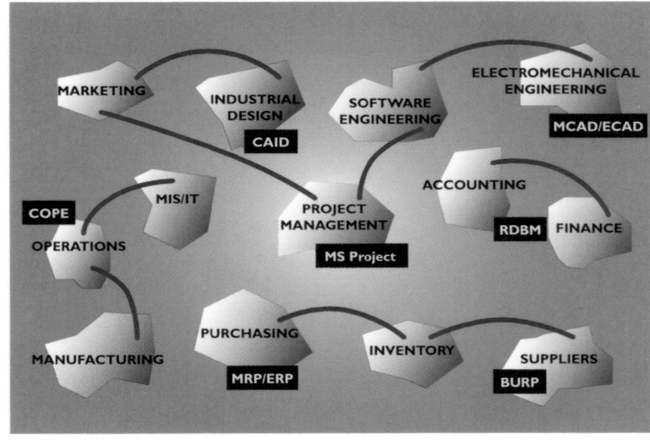

but needing a contiguous "continental" management system.

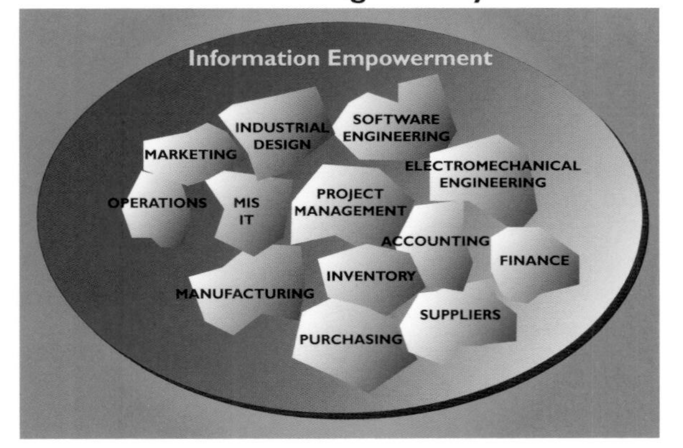

NOTES:

Enterprise Information Process: Critical information management involves capture, control, and accessibility

In most corporate environments, several projects are in various stages of development simultaneously. All project teams must interact with the various functional groups within the corporate enterprise as well as frequently access and exchange critical information among themselves. Information gained from one project that might prove useful to another should be accessible and shared. Various groups that support development must exchange critical information for project as well as infrastructure success. Getting all of the information that is generated into digital format, capturing it into a common corporate reservoir, and making available to everyone benefits the entire enterprise. Universal sharing of information is fundamental to optimizing product development and corporate revenue.

TOOLS, TACTICS & TALENT

> *Bidirectional data conduits for all clients*
> *Rigorous and efficient input/output of data*
> *Adequate access terminals for all clients*
> *Single comprehensive data management system*
> *Integrated hardware and software tools*
> *Competent information facilitators/processors*
> *Attitude of information collaboration by all*

RESULTS & BENEFITS

> *Flexible, easy access via digital format*
> *Optimized overall development process*
> *Improved product quality and profitability*
> *Lower frustration and improved client attitude*
> *True information-based enterprise*

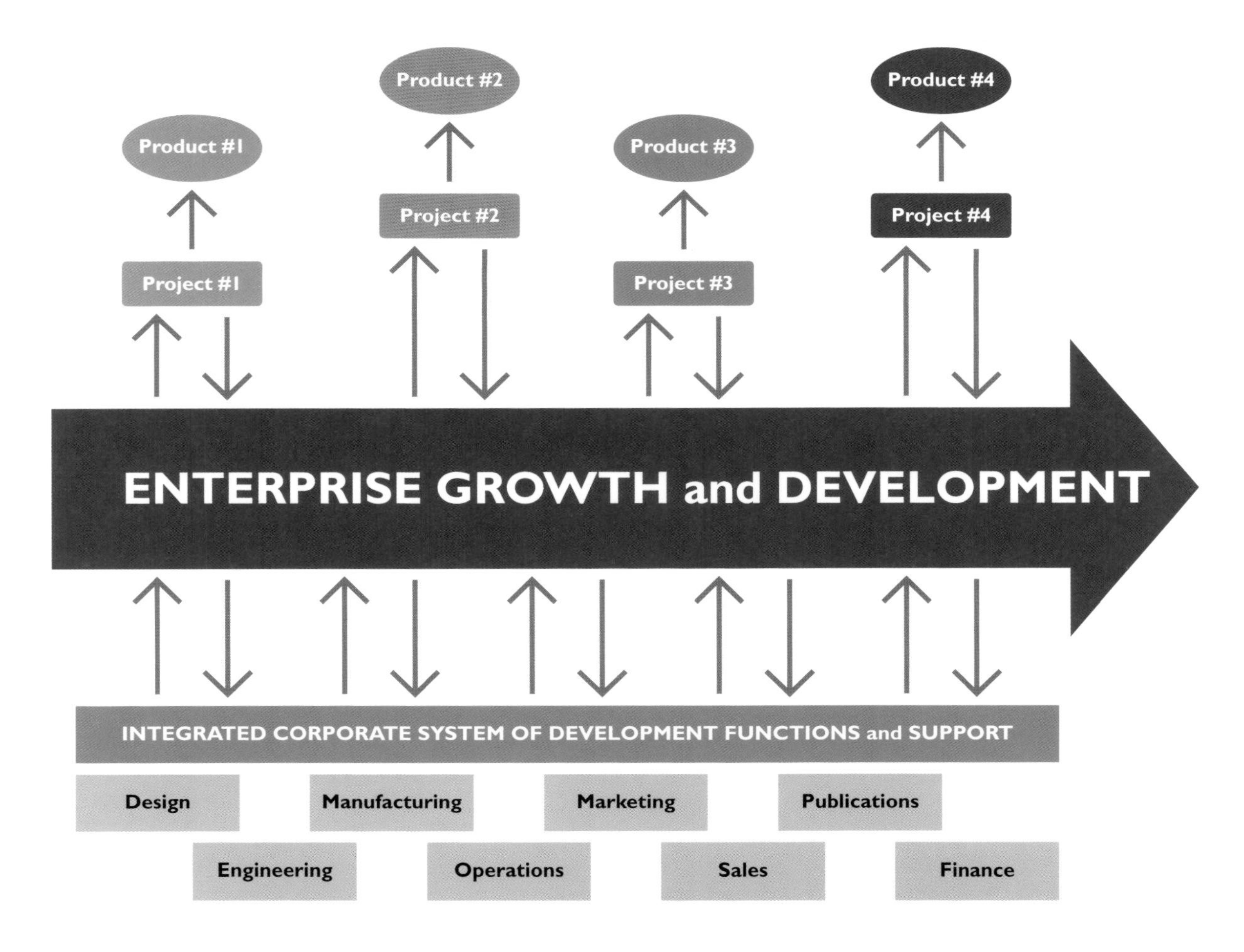

NOTES:

Filling the Information Reservoir: Capturing all information in one location from throughout the enterprise

Diligent capture of information into a single enterprise reservoir is essential to successful information management. This would include not only written, verbal, and alpha-numeric data but also visual information developed from all sources. Development sketches, drawings, mockups, schematics, files, video notes, etc., should be put into the reservoir. Capturing all of this information requires the use of a number of tools and techniques that include audio, video, two- and three-dimensional scanning, digital photography, etc. These capture tools would need to be readily available, easily accessible, and quickly set up. To insure the success of the information reservoir, a process champion should be assigned who provides the vision and energy to optimize this process.

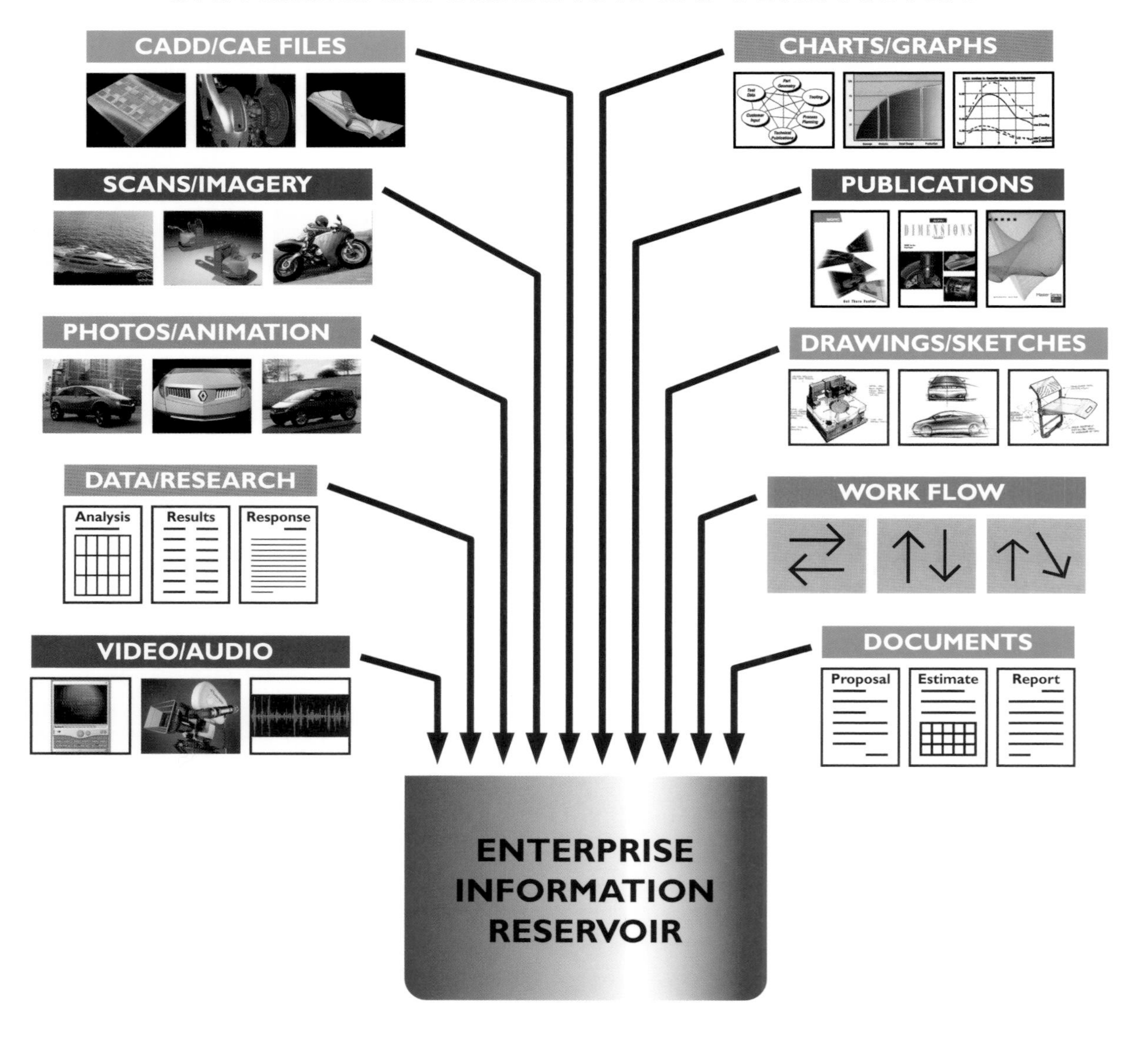

TOOLS, TACTICS & TALENT

- *All information converted to digital format*
- *Assigned champion with vision and motivation*
- *Tools and techniques for digital capture*
- *Information capture from start to finish*
- *Specialists used as required for capture and input*

RESULTS & BENEFITS

- *More time for development innovation*
- *Improved employee productivity*
- *Data and information for future use*
- *Increased collaborative innovation*
- *Improved enterprise profitability*

NOTES:

Tapping the Information Reservoir: Everyone having easy access to information required to do their job

Having an established system for filling the enterprise information reservoir with relevant content is only half the battle. Nothing slows down a problem resolution process more than not having needed information easily and quickly at hand. This is a frustrating and inefficient way to do business and dangerous to success. An integrated system must be implemented for tapping into the reservoir. Retrieval and feedback must be quickly and easily available to every individual in the enterprise, both internally and externally, including outside suppliers. Access to the information must be appropriately secure and on a need-to-know basis. Other than that, there should be free, fast, and user-friendly access. A feedback loop for updating, modifying, and adding to the master database should be integrated into the process.

Tapping the Reservoir to Get Work Done by All Corporate Customers

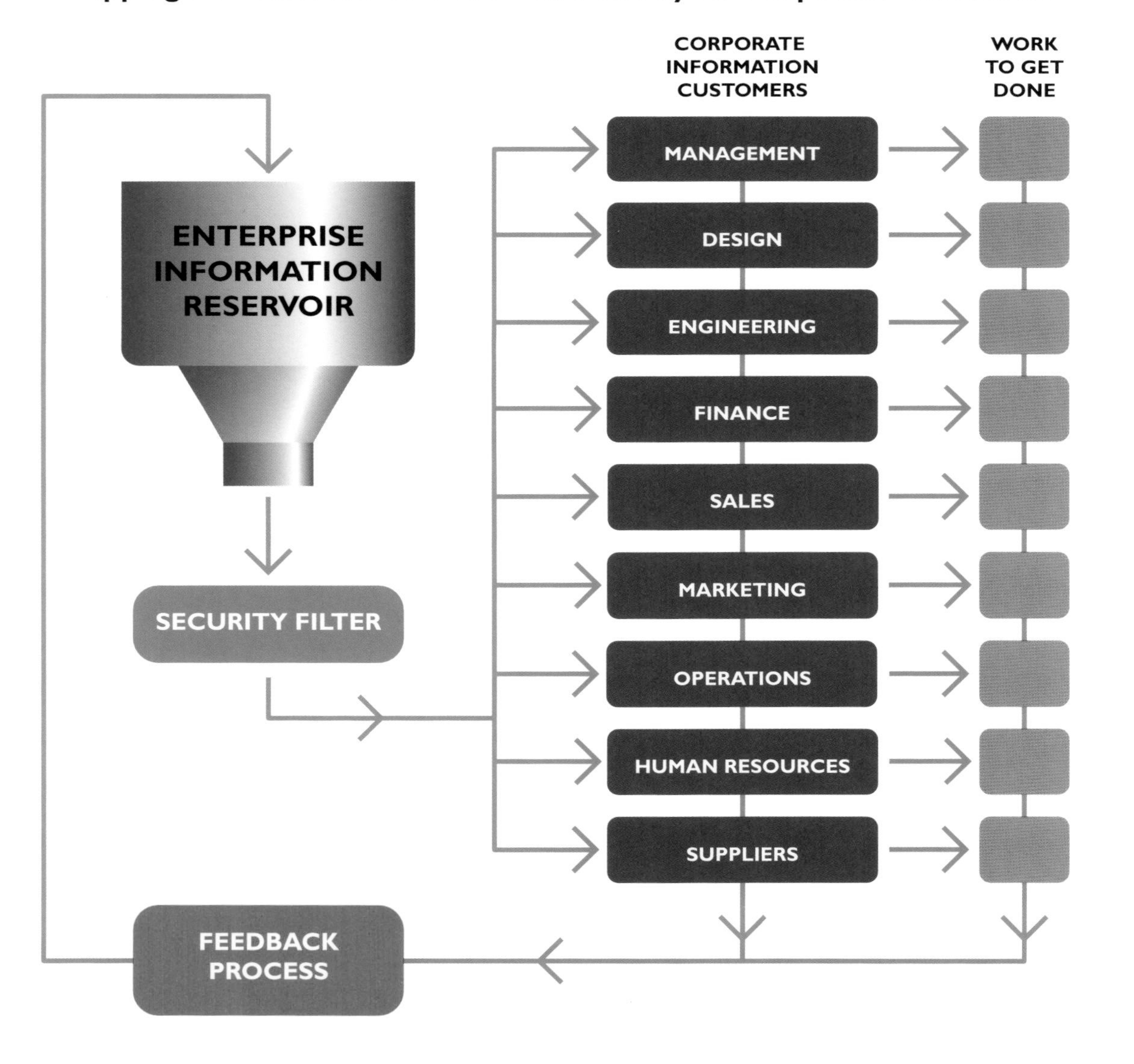

TOOLS, TACTICS & TALENT

- > *Mutual understanding of all functional needs*
- > *Employees as important information customers*
- > *Intolerance to inaccessibility of information*
- > *Process champion to provide vision and support*
- > *Frequent system review and optimization*

RESULTS & BENEFITS

- > *Highly productive and effective employees*
- > *Minimal frustration and maximum satisfaction*
- > *Efficient and streamlined problem solving*
- > *Improved overall bottom line performance*
- > *All enterprise processes optimized*

NOTES:

Information Management Power: Using the proper tools for powerful total information management

One of the most extreme applications for total information management using a single source of data is in the aerospace and aircraft design and manufacturing industries. Boeing Commercial Airplanes Group had previously used multiple data sources and several information management tools to coordinate its vast worldwide reservoirs of information. Over the past several years the company has completely rethought and fundamentally changed their data management philosophy, strategy, and tactics. Through its Define and Control Airplane Configuration/Manufacturing Resource Management initiative (DCAC/MRM), the company has moved to utilizing a single system of tools and tactics for managing a single data source, the ideal scenario for organizing a high volume of complex data and information in a comprehensive and integrated manner across the enterprise. Such an approach is easier to use, simpler to access, and works more accurately in all aspects of information management.

TOOLS, TACTICS & TALENT

> *Use of single information/data source*
> *Simple and accessible input/output process*
> *Internal experts and specialists for training, maintenance, implementation, and support*
> *Senior management buy-in of tools*

RESULTS & BENEFITS

> *Single user interface*
> *All employees have access to information*
> *Everyone gets their job done*
> *Comprehensive management of all information*
> *Assurance of product safety through coordination*
> *Improved financial visibility and performance*

Boeing's Plan for Total Information Management with a Single Data Source

(http://www.boeing.com/commercial/initiatives/DCACMRMOverview/sld001.htm)

The Mission

To fundamentally rethink and radically simplify the processes related to airplane configuration definition and production.

The Approach

Single Source of Product Data	Simplified Configuration Management	Tailored Business Streams	Tailored Materials Management
• Single Bill of Material	• Eliminate Effectivity from Drawings	• TBS 1 Basic & Stable • TBS 2 Previously Delivered Options • TBS 3 New/Unique Options	• Rate Schedule (Very Simple) • Kanban (Simple) • MRP II (Complex)

1

Single Source of Product Data

(Bill of Material)

Engineering BOM
Engineering BOM
Manufacturing BOM
Materiel BOM
Support BOM
Ordering BOM
Planning BOM
Planning BOM
Scheduling BOM
Purchasing BOM
DISCREPANCIES
REPLACED BY
Single BOM

As-Is: Multiple Bills of Material in Separately Entered Legacy Systems

To-Be: Single Bill of Material Entered in Single Source of Product Data

2

Single Source of Product Data

(Information Systems)

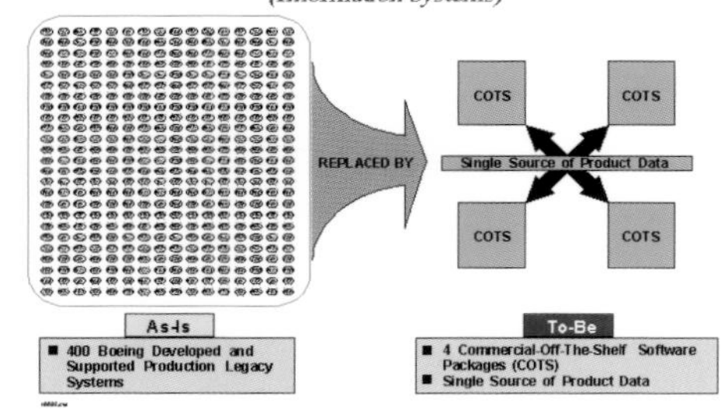

3

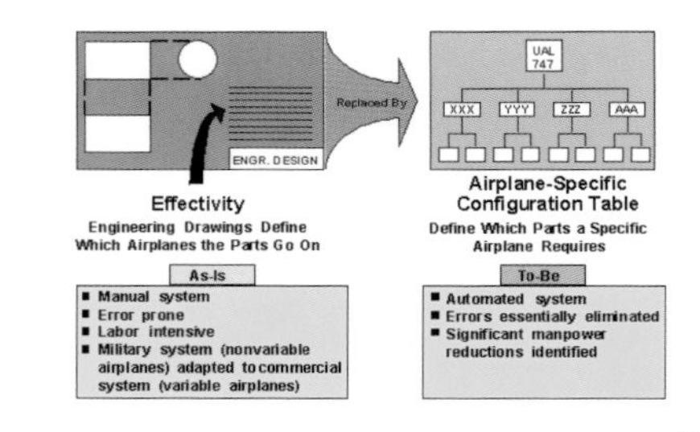

4

Tailored Business Streams

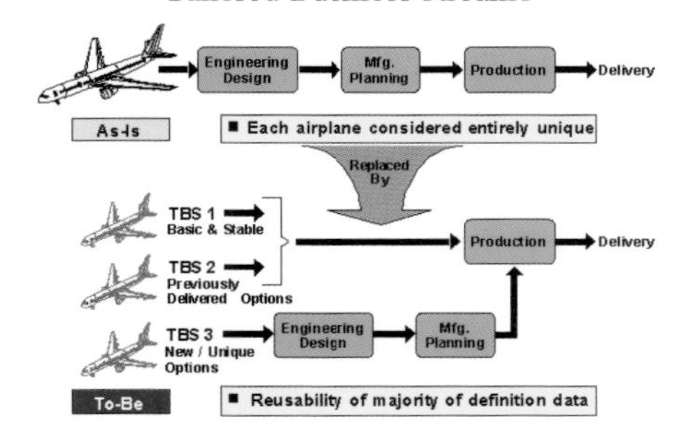

5

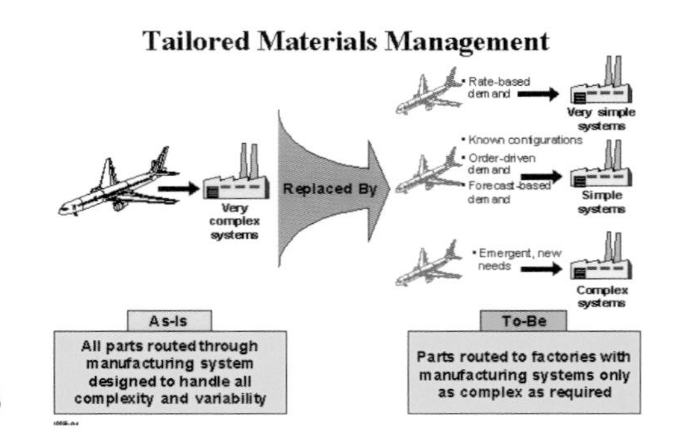

6

NOTES:

Managing Mega-Information: The task of organizing millions of entities for one product line

With its incredible depth and breadth, the Boeing Company must process, access, and control millions of parts, subassemblies, and related documents and information continuously worldwide. This system has been integrated into the design and manufacture of aircraft and into the corporate business structure. Coordinating the company's massive amount of data by integration into a single system has led to a number of significant benefits for this world-class company. These benefits include highly streamlined information processing and accessibility, more efficient and effective product design and manufacturing, increased management oversight of multiple engineering and manufacturing bills of materials, better coordination with outside suppliers, and increased profitability and financial performance.

The Boeing Company: P.O. Box 3707, M/C 10-06, Seattle, Washington 98124-2207

THE BOEING COMPANY

Boeing is the largest aerospace company in the world, the largest manufacturer of commercial jetliners, and the nation's largest NASA contractor. It is also a world leader in the development and production of military aircraft and defense system products and programs. Boeing is organized into four major units: Commercial Airplanes, Space and Communications, Military Aircraft and Missiles, and Shared Services. Boeing and its subsidiaries employ approximately 190,000 people and provide services to customers in 145 countries worldwide.

Acres of production facilities

State-of-the-art transportation products

Massive product assemblies

Large-scale manufacturing operations

NOTES:

"Making the simple complicated is commonplace; making the complicated simple, awesomely simple, that's creativity."
Charles Mingus

CULTURE CONFLICT

The engineering/manufacturing conundrum

Two groups will invariably clash if they have differences in training and education, rewards for their work, goals and objectives for outcomes, management styles, and views of their importance to the development enterprise. Such is often the case with corporate engineering and manufacturing groups. Metaphorically, engineering is chartered to develop an innovative and creative one-off design such as an original painting. Conversely, manufacturing is chartered to efficiently mass produce as many copies of that one-off painting as possible in the shortest period of time. Not only do these objectives often clash, but the transition of getting the one-off engineering design into manufacturing for mass production is too often a painful experience. The solution is mutual understanding, appreciation, and participation throughout a project.

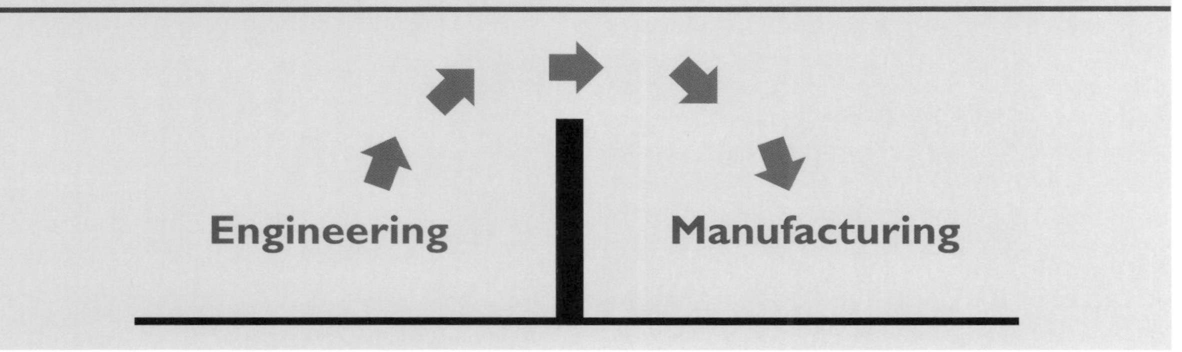

TOOLS, TACTICS & TALENT

- > *Early engagement of engineering and manufacturing on all projects*
- > *Dedicated participation of both sides throughout all projects*
- > *Mutual understanding/appreciation of both sides' objectives and goals*
- > *Development of hand-off management process*

RESULTS & BENEFITS

- > *Smoother, more successful engineering/manufacturing hand-offs*
- > *No more "throw it over the wall" transitions*
- > *Improved professional and personal relationships*
- > *Optimized project schedules, budgets, and quality*
- > *Overall enterprise performance and profitability*

The Engineering/Manufacturing Conundrum

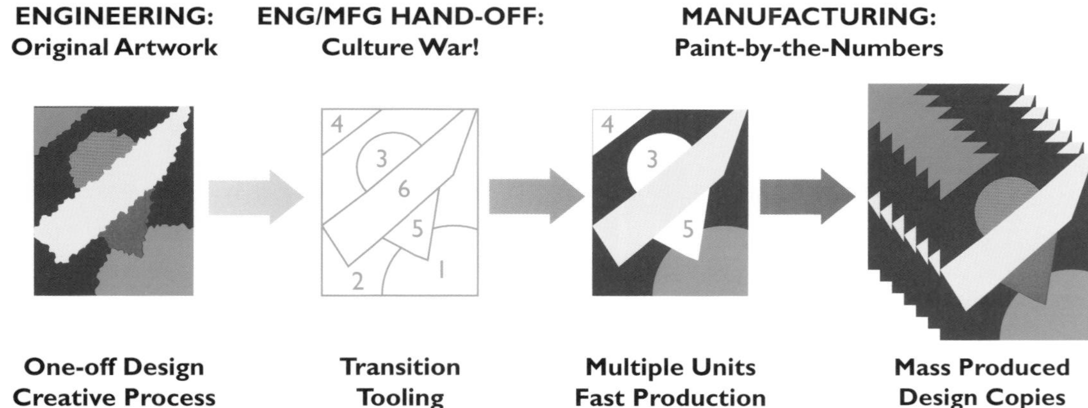

NOTES:

Culture Conflict Solution: Making salad dressing from oil and vinegar

"Mixing engineering and manufacturing people is like mixing oil and vinegar. The problem is that you must make salad dressing!" This astute metaphorical characterization by Craig Polnoff, an experienced designer, engineer, and manufacturing specialist who has worked on both sides, will ring true to those who have ever done product development. Often these two functional groups are separated by department, distance, location, philosophy, budgets, and other attributes that keep them from easily "mixing the salad dressing." But the mixing, as Craig points out, is essential. Getting these two groups to collaborate often throughout a project to jointly meet their discreet functional objectives will only benefit both the project and the company (and make better "salad dressing").

Craig Polnoff: Mechanical HDK Engineer, Palm, Inc., Santa Clara, California

Craig has been an engineer for over fifteen years and worked at several prominent companies while gaining knowledge in product design, supplier quality engineering, and tooling engineering. Craig was a sheet metal fabrication specialist for Hewlett-Packard, Inc., a supply base management engineer at Apple Computer, Inc., and a product design tooling engineer at InFocus Corporation. Craig is currently a mechanical HDK engineer in the continuation engineering department at Palm, Inc., where he emphasizes the importance of details and quality when manufacturing products.

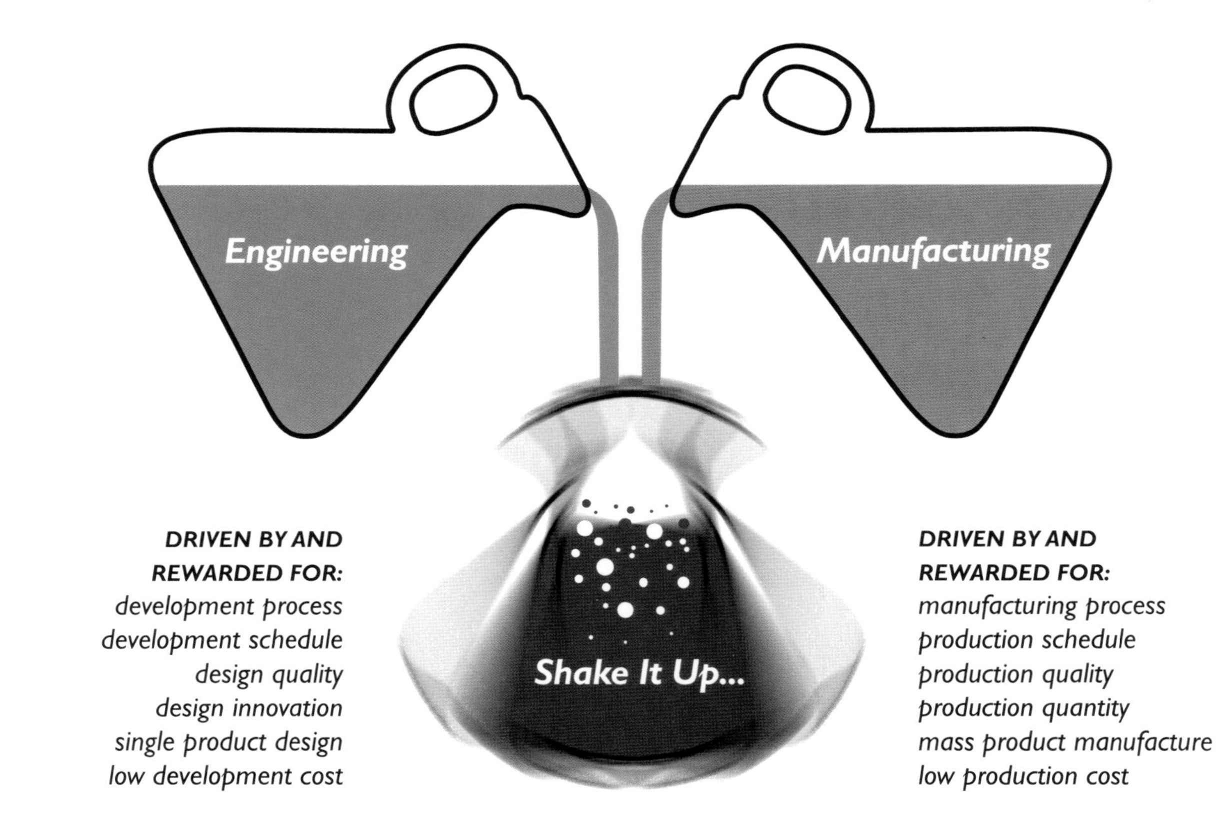

NOTES:

"Fail often in order to succeed sooner."

Motto at IDEO

RPM OPTIONS

Making appropriate prototyping decisions regarding materials and processes for optimum performance

There are many RPM options in materials and processes available to the aggressive development company, and these options are constantly changing and expanding. Each carries certain attributes relative to quality, speed, and cost. Decisions must be made as to which process should be used for a specific application in a development project. Questions to ask include: Is this a conceptual or detailed prototype? How robust or durable does it need to be? Is this prototype for form/fit evaluation only or must it withstand test stress loads? How closely must this prototype reflect the final part finish and material? The chart on this page reflects only a few processes and a moment in time but provides an idea of the variety of processes available with their rough features and attributes. Such a chart must be constantly updated to provide appropriate information for the RPM decision-maker.

TOOLS, TACTICS & TALENT

> *Dedicated RPM implementation specialist*
> *Constant state-of-the-art research*
> *Accessibility and communication of processes*
> *Development of suppliers and resources*
> *Proper application of processes and materials*
> *Understanding of advantages/limitations of tools*

RESULTS & BENEFITS

> *Optimized product development via prototyping*
> *Reduced costs via appropriate applications*
> *Optimized internal versus external RPM suppliers*
> *Improved design visibility via prototypes*
> *Early optimized product testing via prototypes*
> *Better design and bottom line results*

The chart below represents a very qualitative evaluation guide for selecting a number of currently mainstream RPM processes applicable to product design and development needs for producing a moderately complex part. The data was developed with Jim Burt of InFocus Corporation and Michael Duncan of ARRK Product Development Group. However, any errors or misjudgments are the author's.

CHOOSING THE RIGHT RPM PROCESS AND MATERIAL

CATEGORY	CNC	SLA	FDM	SLS	SRM	3DP	RPT	CPT
Tooling or setup cost	low	low	low	low	low	low	med	high
Execution speed	med	high	med	high	med	high	med	low
Part cost	med	med	med	med	med	low	low	low
Part durability	high	med	high	med	med	low	high	high
Part precision	high	med	med	med	med	low	high	high
Part quantity	low	low	low	low	med	low	high	high
Finish quality	high	med	med	med	high	med	high	high
Shippable parts	(yes)	no	no	no	no	no	yes	yes

LEGEND

CNC = Computer Numerical Control: precision machined parts of metal or plastic
SLA = StereoLithography Apparatus: proprietary laser-cured resin parts
FDM = Fused Deposition Modeling: sequentially layered thermoplastic parts
SLS = Selective Laser Sintering: sequentially layered powdered resin parts
SRM = Silicon Rubber Molding (RTV): machined pattern to rubber mold to cast urethane parts
3DP = 3 Dimensional Printing: sequentially layered waxy resin parts
RPT = Rapid Prototype Tooling: quickly made epoxy or metal molds to molded parts
CPT = Conventional Production Tooling: robust metal injection tooling to molded parts

low = qualitatively low, few, or slow
med = qualitatively medium, moderate, or middle
high = qualitatively high, many, or fast

(For further RPM process and supplier information go to: **www.cc.utah.edu/~asn8200/rapid.html#COM***)*

066

NOTES:

RPM Champion: An aggressive program using a variety of processes pays off in speed and quality

As with any critical process that is significantly important to the product development enterprise, that enterprise must both aggressively pursue constant state-of-the-art status as well as retain the appropriate talent to expedite it. InFocus does both for their RPM components of product development. Jim Burt is the manager responsible for providing expertise, tools, materials, resources and suppliers for RPM efforts. Jim manages, facilitates, and/or directs both in-house resources as well as outside suppliers. He champions state-of-the-art tools, tactics, and talent to senior management to support the engineers and designers needing the best resources possible for their development work. This page features a number of examples of applications that Jim and others integrate into their product development projects in support of InFocus' aggressive time-to-market and profitability objectives.

Jim Burt: Product Design Services Manager, InFocus Corporation, Wilsonville, Oregon

Jim comes from a background of both contract and corporate engineering in product development. He has been active in product design, machine design, mechanical design, and manufacturing facilitation and fabrication. Originally hired as a design engineer at InFocus out of consulting, he now successfully manages several functional support groups within the company including RPM processes, MCAD documentation and CAE technology support. Jim's passion is to optimize any and all development processes with state-of-the-art technology and talent so as to maximize company profits.

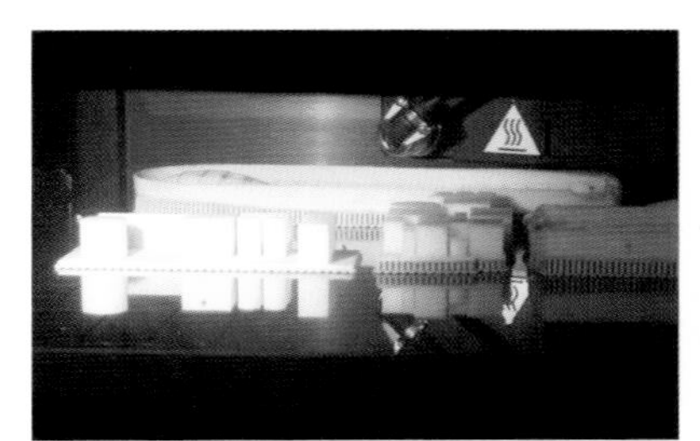

InFocus in-house FDM system making a part

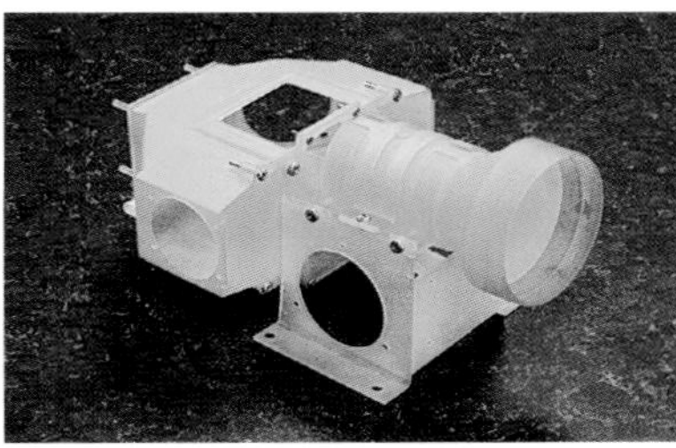

Prototype lens support assembly in SLA

SLS lamp housings for thermal testing

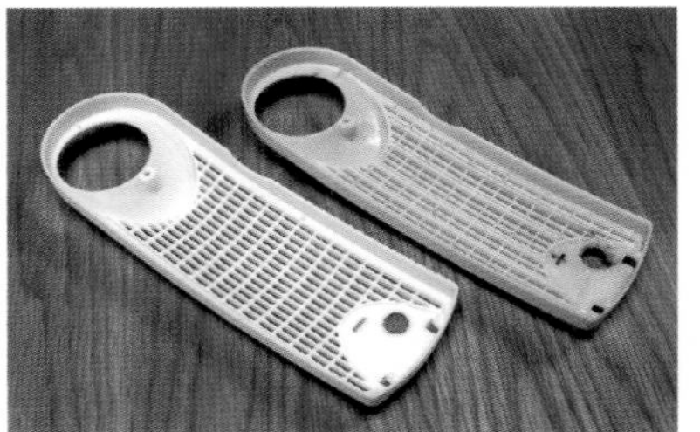

Rubber-molded and rapid-tooled bezels

Precision rubber-plaster casting prototype

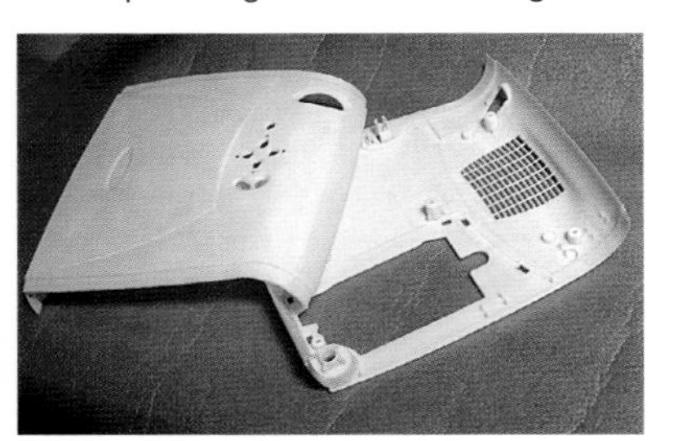

Functional and robust FDM enclosure covers

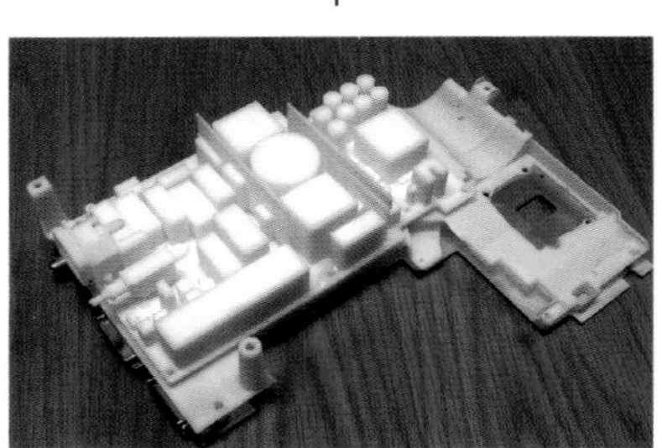

PCB component space mockup in FDM

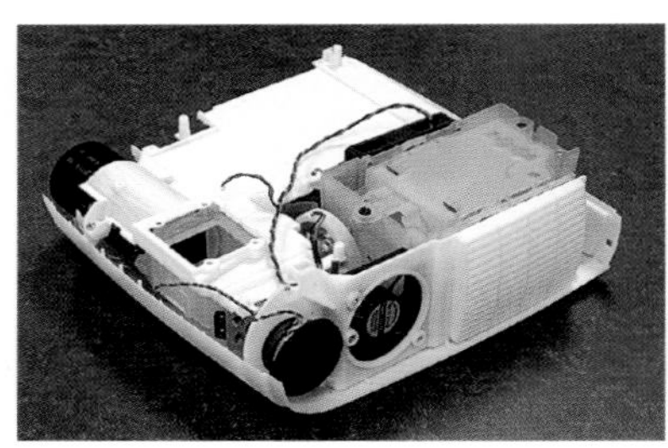

System assembly in mixed prototype media

Clear and CNC LP720 prototype test units

TOOLS, TACTICS & TALENT

- *Champion RPM facilitator/manager/proponent*
- *Appropriate in-house resources based on ROI*
- *Positive relationships with best outside suppliers*
- *Constant research/development of latest TT&T*
- *If in doubt, prototype!*

RESULTS & BENEFITS

- *Large cost savings with proper in-house processes*
- *Early prototyping speeds development process*
- *RPM adds unique 3D and kinesthetic look/feel*
- *ROI of RPM pays off in profitability and TTM*
- *Prototypes provide visibility to others in company*
- *RPM investment benefits speed and quality*

NOTES:

Early Physical Mockups: These get issues resolved quickly when facilitated fast and early in a project

Even with the incredible virtual simulation and digital imagery technology available to designers today, the need for actual physical models that can be viewed and kinesthetically manipulated still has great advantage and value. Physically handling an object has its own unique benefits over viewing it on a computer screen. This is why RKS Design has developed its system of rapid modeling of product concepts in their design shop. These quickly executed foam mockups with varying levels of detail are produced via CNC machining and are produced directly from digital computer models. The fast production of a physical entity enhances the client s response, evaluation, and interactivity very early in the development process before large expenditures have been committed. Thus, early physical evaluation speeds closure of the design process and reduces development cost.

TOOLS, TACTICS & TALENT

> *Adequate CNC hardware and software*
> *Talented staff of model makers*
> *Adequate shop space and facilities*
> *Champion committed to aggressive facilitation*
> *Utilization of process throughout all projects*

RESULTS & BENEFITS

> *Early concept exposure to client*
> *Early client evaluation and choices*
> *Fewer surprises and conflicts later*
> *Faster design concept resolution*
> *More productive and faster schedule*

Juan Cilia: Shop Manager and Model Maker, RKS Design, Inc., Thousand Oaks, California

Juan manages RKS Design s model shop and coordinates with the design studio for the transfer of files to the CNC machines for prototyping. With three CNC machines, RKS is able to produce superior models for use in exhibits, client sales presentations, and as engineering and manufacturing guides. These models give clients an accurate representation of the "final product." Juan holds a B.S. in industrial design and is skilled in Surfcam, a specialized computer aided machining software. His past experience as a contractor includes affiliation with companies such as Mattel and Rubbermaid as well as movie and television studios and other design firms.

CAID-generated virtual models

Extensive shop facilities

Multiple-process interaction

CNC implementation

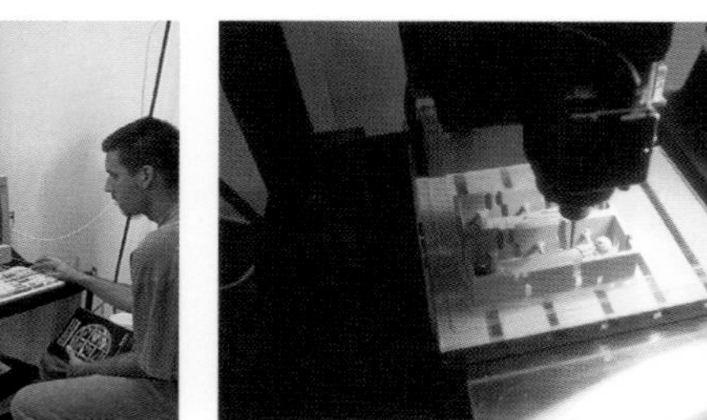

Precision model machining

Rough concept machining

Multiple levels of models

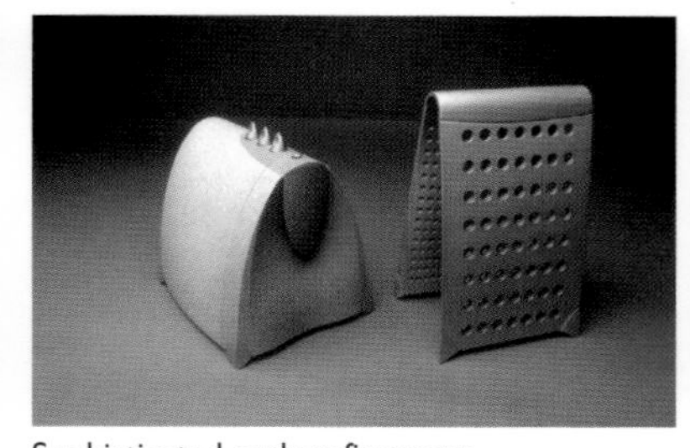

Sophisticated early refinement

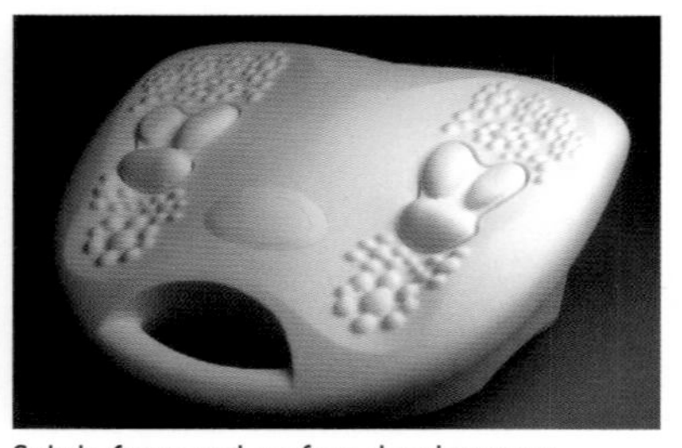

Subtle form and surface development

NOTES:

No More Paper: Eliminating fully dimensioned paper engineering documentation

With the use of CAID and CAE software tools and RPM technologies, it is rare that the progressive, fast-paced company has a need for fully dimensioned engineering paper drawings for a product system. Even though the latest MCAD software will produce such drawings quickly and easily, they are seldom needed for tooling and manufacturing as nearly all data is now communicated digitally. Dimensions, geometry, features, specifications, tolerances, materials, etc., can all be communicated directly via the comprehensive CAD model file. Tooling fabrication, image communication, specification, component ordering, etc., are facilitated digitally, often via an Internet FTP site or network system. Digital or paper drawings are made, if at all, simply to communicate abbreviated checkpoint and quality control information. The era of the tedious design-draw-print-check-revise-print-check-release-file process is over.

TOOLS, TACTICS & TALENT

> *Comprehensive integrated CAE tools*
> *Integration of all specifications in CAE files*
> *Implementation of single PDM system*
> *Elimination of paper-based documentation*
> *Implementation of electronic documentation*
> *Electronic manufacturing facilitation*
> *Development information accessible to all*

RESULTS & BENEFITS

> *Efficient electronic documentation system*
> *Universal information access at all times*
> *Faster documentation/specification process*
> *Optimized engineering to manufacturing hand-off*
> *Shorter project schedules and time-to-market*
> *Improved project profitability and performance*

Marty Maiers: Senior Product Design Engineer, PCS Scanning, Inc., Eugene, Oregon

Marty has worked in the high-technology industry for twenty-one years in corporate and consultant venues. He currently is both an independent consultant and design engineer at PSC Corporation where he manages design teams that continually advance laser scanner technology. Before working at PSC, Marty was a principal product design engineer at InFocus Corporation, a product design engineer at Spectra Physics, and a product design engineer at Victor/NNA, where he helped design the first European desktop computer. Marty believes that as a designer his most important objective is not to deliver a paper documentation package but to help deliver tooling that can be used to produce revenue-generating products.

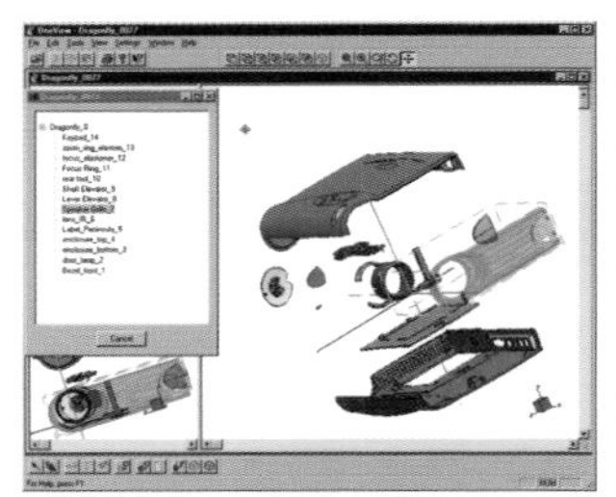

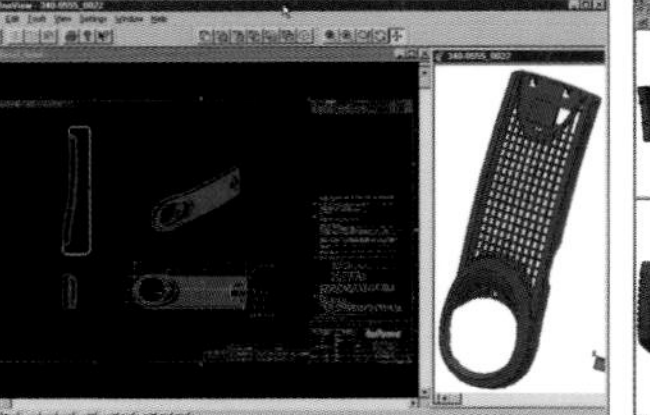
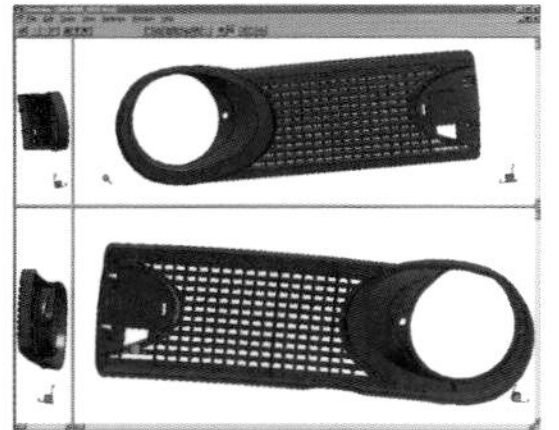

Comprehensive design and manufacturing data resident in MCAD model files and digitally accessible to all

NOTES:

> "People cannot discover new lands until they have the courage to lose sight of the shore."
>
> Andre Gide

RPM ALTERNATIVE

An important option for decreasing time-to-market

Product time-to-market is crucial to market position, brand leadership, revenue generation, and general corporate success. Releasing a product even one month ahead of schedule, or prior to the competition's introduction, can be key in determining market share and profitability. However, shaving off a month in today's aggressive high-technology product development cycles is extremely difficult. This is especially true when using conventional manufacturing processes. Though these processes are generally necessary for long-term robust productivity and cost savings, they are inherently slow compared to the needs of today's time-to-market objectives. With new rapid tooling technologies, an interim process can be used to speed time-to-market while the robust systems are being developed. This process involves the concurrent development of rapid tooling that, though short-lived and with higher part cost, will get product to market quickly while the robust system is phased in.

TOOLS, TACTICS & TALENT

- *Dedicated and distinct RPM implementation team*
- *Top management buy-in of concurrent approach*
- *Focused and dedicated RPM suppliers*
- *Buy-in by operations team of dual paths*
- *Feedback of RPM data/results into robust system*
- *Champion leader for RPM alternative approach*

RESULTS & BENEFITS

- *Optimized robust path from RPM feedback*
- *Significant time-to-market advantage*
- *Increased revenue and market share*
- *Potential first-to-market advantage*
- *Earlier market response and feedback*
- *Improved overall project ROI*

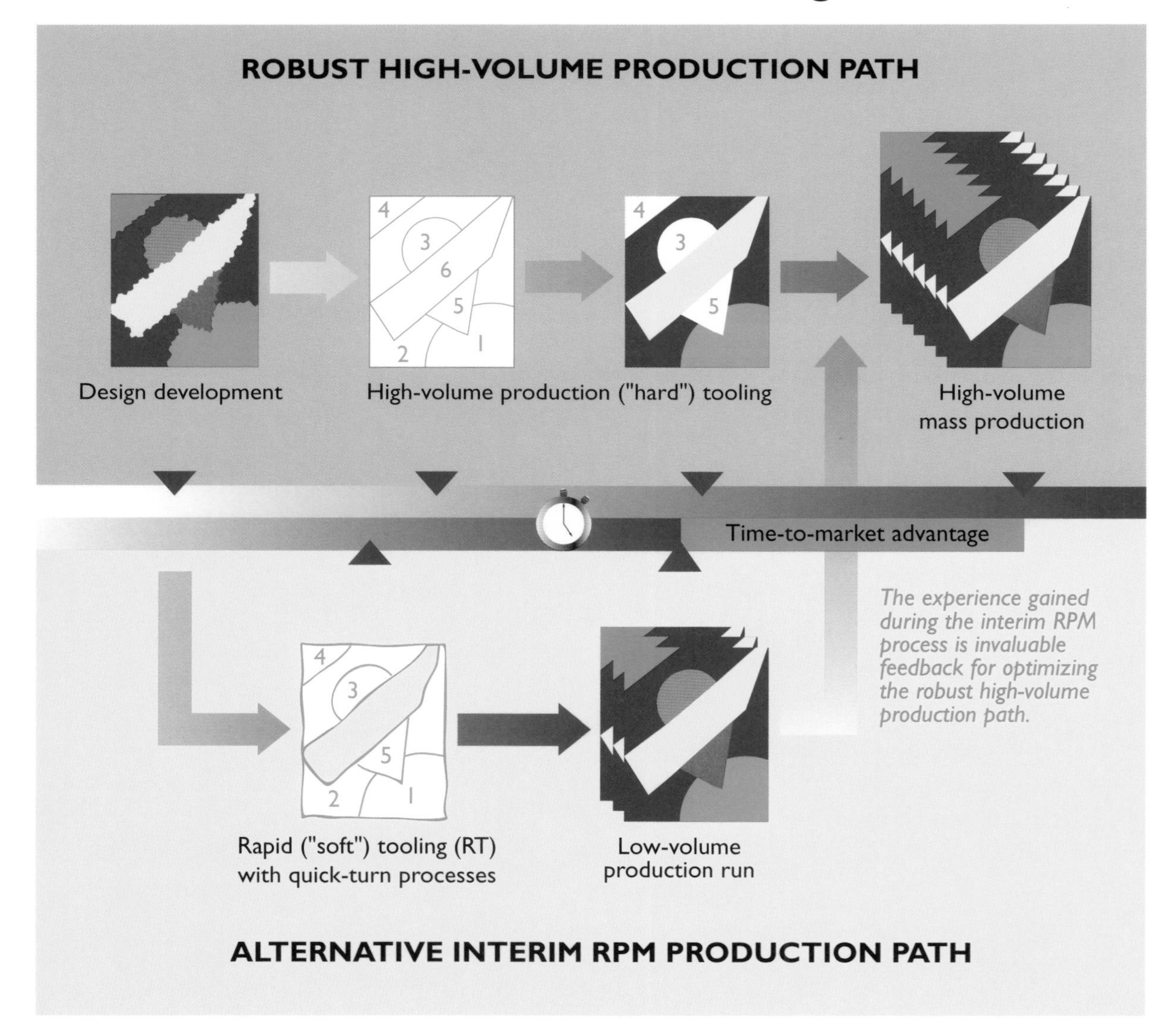

NOTES:

RPM in Action: A broad range of applications can improve time-to-market, profitability, and market share

A number of both independent and internal RPM organizations serve the product development community. The best of them provide state-of-the-art rapid prototyping and rapid tooling/manufacturing capabilities to speed the development process and win the time-to-market race. Many offer a broad range of RPM resources such as appearance modeling, CNC machining, urethane and epoxy casting, SLA and SLS prototypes, rapid soft and hard tooling, injection molding, finishing, assembly, and other short-run and production services. ARRK Product Development is such a company, having several sites worldwide and many of the above services to assist their customers in introducing new products quickly into the marketplace. More often than not, such services provide the cutting edge required to meet or beat aggressive project schedule targets. In some cases the RPM solution will be sufficient for the entire production run of a product. Such time-to-market advantages can result in significant corporate profitability and market share.

Michael Duncan: Northwest Regional Manager, ARRK Product Development, Beaverton, Oregon

Michael's passion is to provide the best prototype and tooling solutions while continually improving the speed and quality of new product introduction. He is the business development manager for ARRK Product Development Group and also supports the manufacturing aspect of new product development as a consultant. Michael has a B.A. in industrial design and holds three patents. He has innovated new processes at Art Center College of Design in Pasadena, California by assisting in introducing the use of SLA and SLS materials and techniques for the industrial design process.

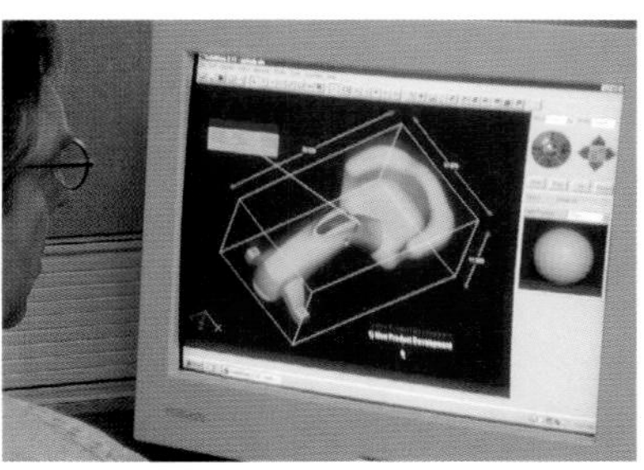

CAE models converted to CAM

Set of rapid tools for plastic parts

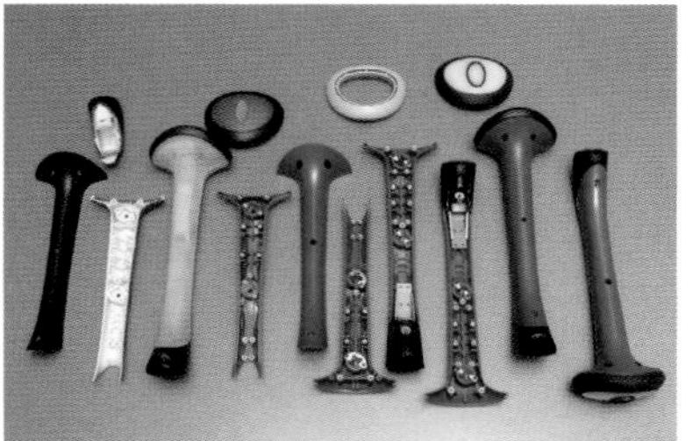

Sequence of RPM parts for a wand device

RPM IMPLEMENTATION STRATEGIES

Concurrent RPM new product introduction (NPI)

Conventional NPI

RPM NPI

Delayed conventional NPI

Conventional NPI

RPM NPI

Market test NPI

Conventional NPI

RPM NPI

RPM NPI only

RPM NPI

Time

TIME-TO-MARKET ADVANTAGE

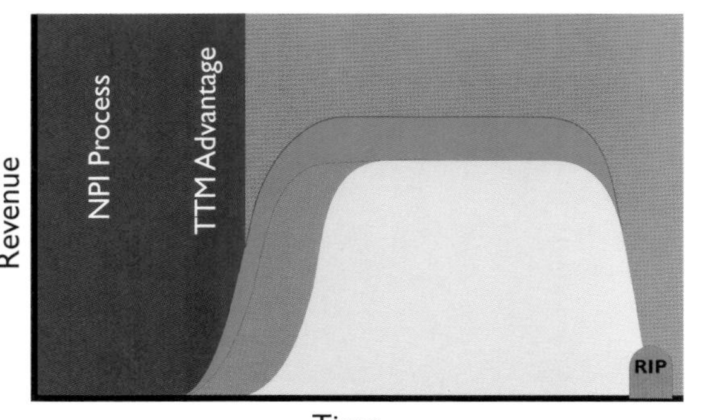

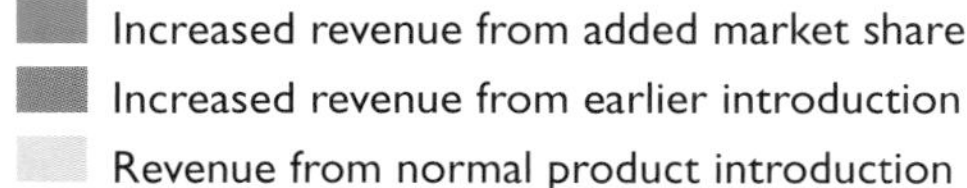

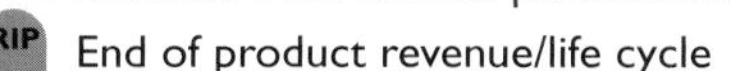

NOTES:

"*Even if you are on the right track, you will get run over if you just sit there.*"

Will Rogers

REVERSE ENGINEERING

From part to art and back again several times

Even with today's amazing digital tools for surface and form development where nearly any shape can be produced with computer technology, there is still a place for the subtleties of hand craftsmanship. Automotive clay modelers know this well: sometimes the most subtle, beautiful, tactile details of a surface or form, whether for appearance or feel, can only be accomplished by hand craftsmanship. With the advancement of three-dimensional scanning technology, the flexibility and efficiency of digital surface development can be combined with the nuances of hand craftsmanship in an iterative process that optimizes look and feel. By iterating between digital models and handcrafted adjustments, a form can be generated, fine tuned, and finalized with speed and efficiency into final digital format.

THE REVERSE ENGINEERING PROCESS

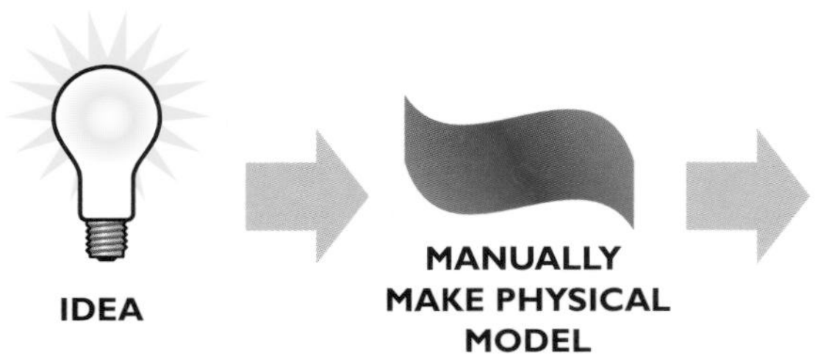

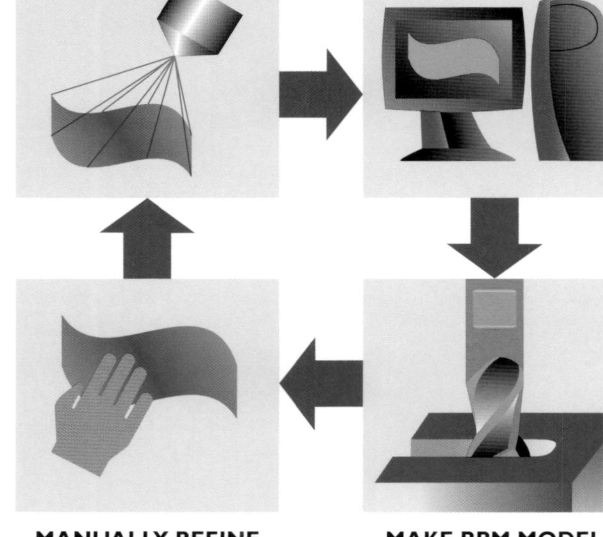

A combination of manual physical modeling, laser scanning, RPM, and virtual modeling that retains the original subtle design intent.

APPROVED

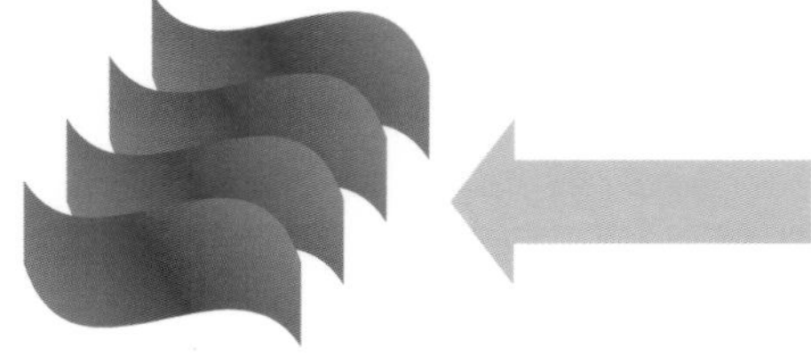

TOOLS, TACTICS & TALENT

> *Talented manual model makers*
> *Talented digital surface developers*
> *Three-dimensional scanning resources*
> *Sophisticated surface development software*
> *RPM prototyping resources*

RESULTS & BENEFITS

> *Finely refined and tuned forms*
> *Optimized touch and feel to products*
> *Often faster surface development*
> *Tactile and kinesthetic interaction of users*
> *Superior sophisticated product features*

072

NOTES:

Reverse Engineering in Action: Creating style with functional human factors by integrating processes

When designing a product that has a significant physical human interface, it is often more effective to start with a physical three-dimensional form-fit model than with a virtual computer-based version. Initial physical modeling helps insure proper human interface conformance. This physical 3D model can then be converted to a digital computer solid model via scanning or dimensional analysis and subsequently refined to the final design, ready for production. Such was the case for Jonathan Dry and the engineering design for a Designworks/USA client's goggles project. First, a master form model was made in modeling foam in the shop by the industrial designer that both aesthetically and ergonomically fit the human requirements of basic form, fit, and function. The dimensions of this model were then converted into an MCAD solid model where the final design details and nuances were refined to completion and ready for tooling.

Jonathan Dry: Project Engineer, Designworks/USA (BMW), Newbury Park, California

Jonathan was the project engineer on a multi-disciplinary team that developed the Altitude goggles for Scott USA. Jonathan's special design interests lie in preserving design intent from concept through manufacturing. As an experienced mechanical designer, he has spent many years becoming proficient in sophisticated MCAD software and produces virtual product engineering models from designers' concepts and carries them through to manufacturing. Jonathan received his B.S. in mechanical engineering from California Polytechnic State University at San Luis Obispo, California.

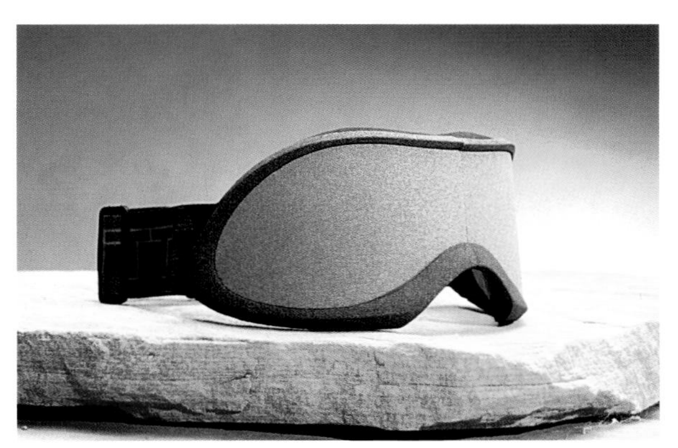

Original foam master model

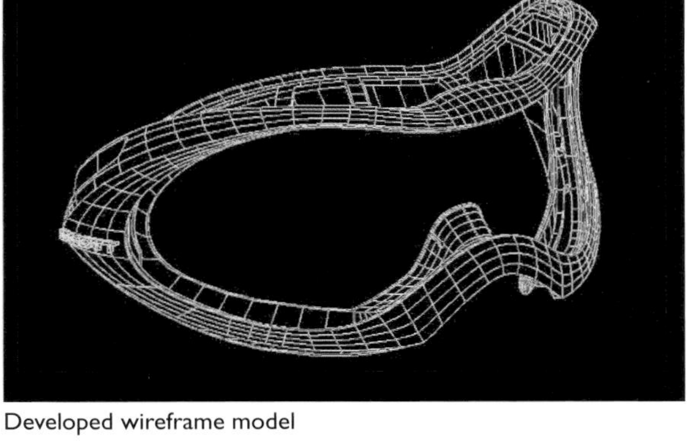

Developed wireframe model

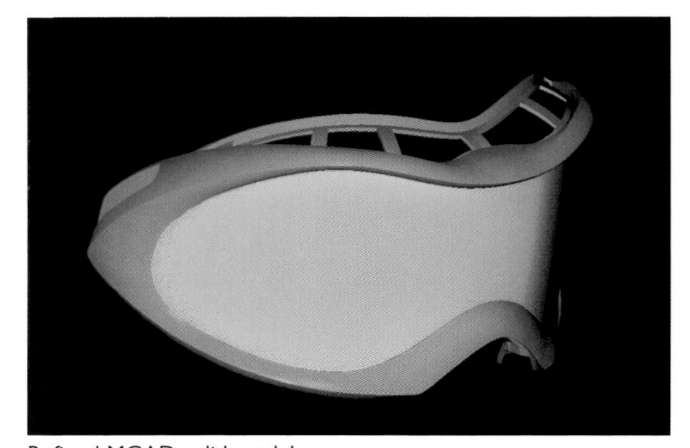

Refined MCAD solid model

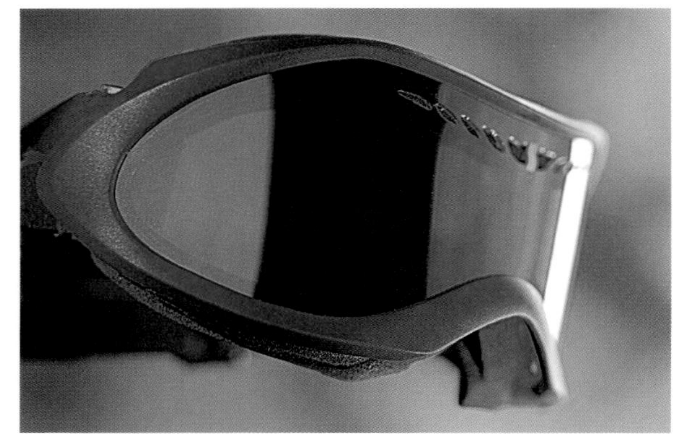

Final production product

NOTES:

"Quality is conformance to requirements, not goodness."
Phil Crosby

CULTURE OF RELATIONSHIPS

Key to a successful, long-term quality design and manufacturing environment

Critical to success in high-technology product manufacturing is the relationship between client and supplier. The best of these relationships are often highly personal. Product component tooling and manufacture is as much a human art and craft as it is a science, and those who facilitate such efforts must appreciate this. Profitable manufacturing relationships need intense, close, aggressive, and intimate communication. This cannot be accomplished solely with paperwork, quotations, RFQs, corporate quality rules, purchasing guidelines, or other routine operations tools. The more aggressive the requirements, the more positive and intimate the relationships must be. The manufacturing facilitator, often a tooling or manufacturing engineer, is essential. The competent interaction of this person between OEM client and suppliers, whether internal or external, is critical to the success of any project.

John Loewen: Senior Product Design Tooling Engineer, InFocus Corporation, Wilsonville, Oregon

John is passionate about getting tooling suppliers to continually improve lead times for faster time-to-market. He is able to promote this conviction in his job at InFocus, since he works closely with product designers and engineers to optimize part design for manufacturing and tooling. He also manages tooling packages for injection-molded plastics, sheet metal, die cast, and thixomolding processes. John received his B.S. in manufacturing engineering technology from Western Washington University and, in addition to his current position, he has served as a senior manufacturing engineer, sales engineer, and procurement engineer.

TOOLS, TACTICS & TALENT

> *Competent manufacturing facilitator*
> *High mutual understanding and trust*
> *Effective communication tools and processes*
> *Compatible software and hardware tools*
> *Minimum paperwork and maximum interactivity*
> *Personal relationships at all levels*

RESULTS & BENEFITS

> *Faster time-to-market production processes*
> *Lowered manufacturing and production costs*
> *Stable and profitable manufacturing partnerships*
> *Fewer disruptions and changes in suppliers*
> *Focused efforts on details, quality, and productivity*
> *Motivated and energized operations staff*

THE MANUFACTURING SUCCESS TRIAD

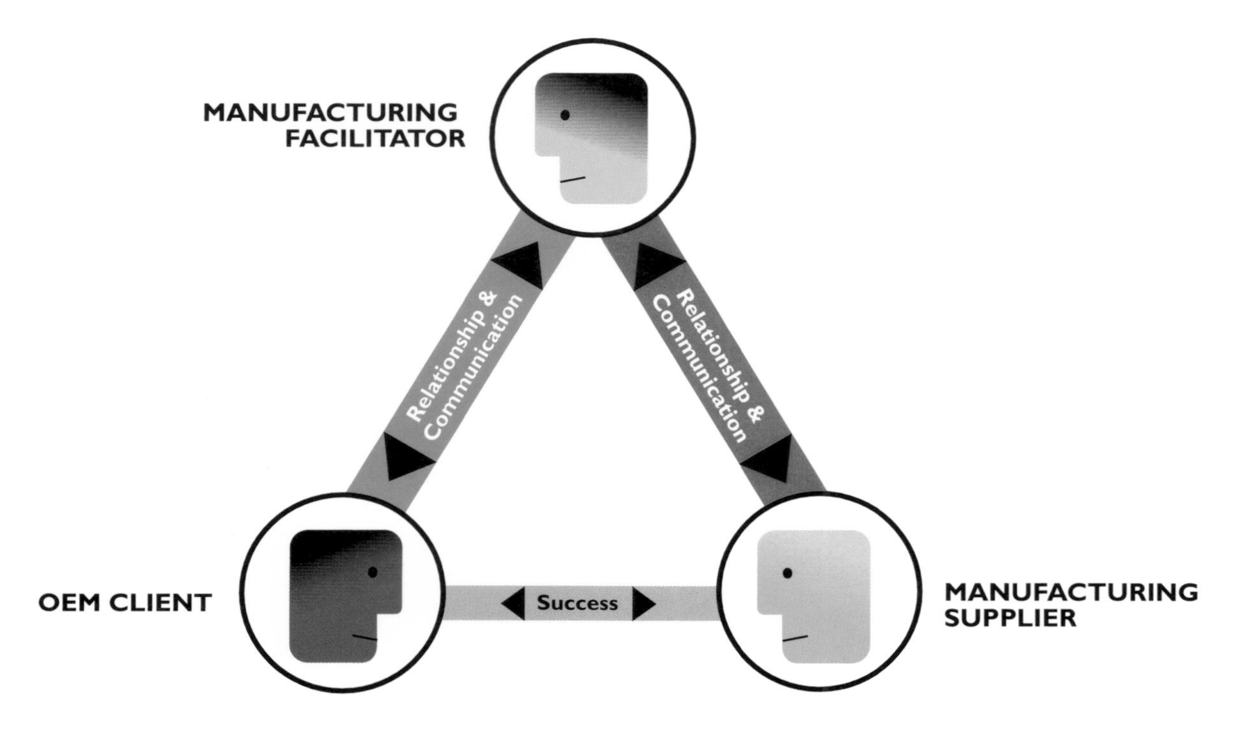

NOTES:

Support Relationships: Critical development functions essential to engineering success

In any high-technology enterprise, design support activities are essential to development success. Critical issues such as thermal performance, electromagnetic radiation control, environmental compliance, safety certification, shock and vibration survival, FDA approval for medical products, materials and components testing and evaluation, etc., must all be resolved for each product or system. To support design and determine compliance, each of these activities often requires at least one individual with specialized tools, if not an entire team with a fully equipped laboratory. At Lexmark, in developing their world-class desktop printers, the product engineering organization has a number of such functional support groups that its designers and engineers can utilize during the development process. Lexmark has organized these services into a system that smoothly integrates into the development process at the appropriate times to insure optimum product performance.

TOOLS, TACTICS & TALENT

- > *Identification of support functions required*
- > *Appropriate and competent personnel*
- > *Adequate shops and laboratories for application*
- > *State-of-the-art instruments, tools, and processes*
- > *Support management for all projects*
- > *Coordination, collaboration, and cooperation*

RESULTS & BENEFITS

- > *Early debugging of critical product issues*
- > *Proper compliance of products to regulations*
- > *Reduced post-manufacturing recalls*
- > *Competent pool of accumulated expertise*
- > *Uniformity of product performance and quality*
- > *Competitive market performance advantage*

Jason Ivy: Product Manufacturing Engineering Manager, Lexmark International, Inc., Lexington, Kentucky

As the product manufacturing engineering manager for Lexmark, Jason Ivy supervises a product manufacturing engineering group that aggressively supports the development of innovative business printers. Jason began working for Lexmark in 1992 and has a passion for the importance of manufacturing engineering and development support functions. He has experience in the manufacturing of a variety of products including keyboards, notebook computers, and laser printers. Jason received a B.S. in electrical engineering and also has an M.B.A.

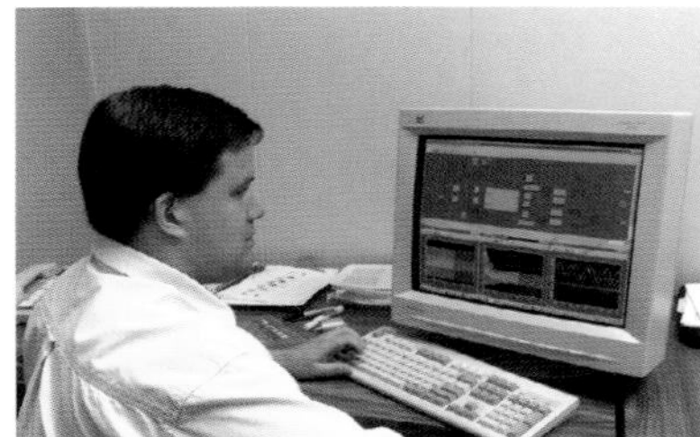

Frequency imaging and analysis

Air flow analysis

Printing precision evaluation

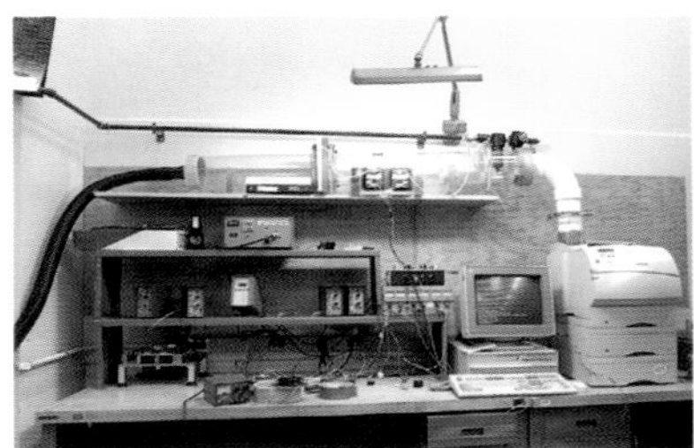

Air particulate testing

Noise suppression testing

EMC testing and compliance

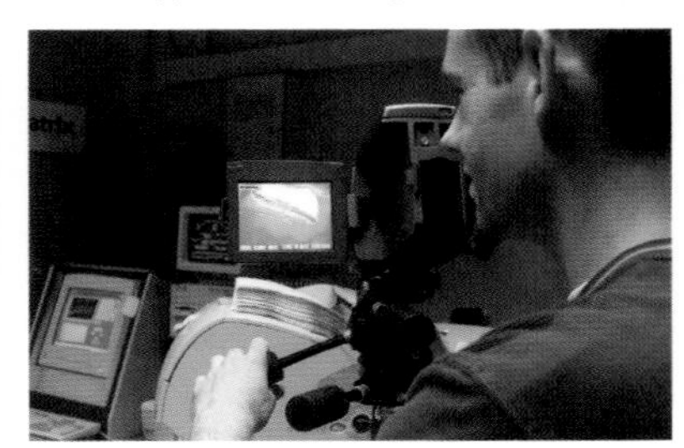

Thermal scanning and mapping

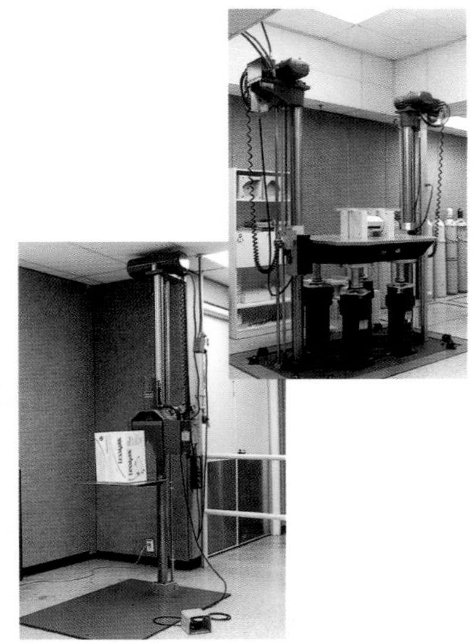

Drop, shock, and vibration testing

NOTES:

"It doesn't matter what product or service you're offering; there is unlimited ability to improve the quality of anything."

Tom Peters

VIRTUAL EVALUATION

Optimizing quality and performance with integrated virtual technology

One of the most tedious processes in product development is dimensional evaluation of mechanical parts and assemblies. Prior to the advent of computer-aided systems, dimensional validation required measuring of a limited number of physical features with manual measuring tools and tedious comparison with dimensions on paper drawings. Even with the advent of computer-based four-axis probe systems, the process is still limited in scope and speed. However, the latest scanning-based dimensional analysis technologies not only drastically reduce the need for direct physical measurement, freeing the engineer for more productive work, but also increase speed, scope, and problem-solving capability. Entire parts and assemblies can be scanned and compared against original design intent with design responses made in a timely manner, since dimensional trends and deviations are visualized and evaluated quickly.

Rautenbach Aluminium-Technologie GmbH: Giesserweg 10, 38855 Wernigerode, Germany

RAUTENBACH

Rautenbach's commitment to the continuing innovation of precision parts has been a tradition for over 100 years. Each year it produces more than 1.2 million cylinder heads for the European automobile industry. Rautenbach has established a personal relationship with many prominent vehicle manufacturers and often takes automobile components from conceptualization to series production. Rautenbach emphasizes the importance of efficient processes and process safety. Both Rautenbach employees and clients benefit from the close cooperation the company has with universities, technical institutes, and well-known engineering businesses.

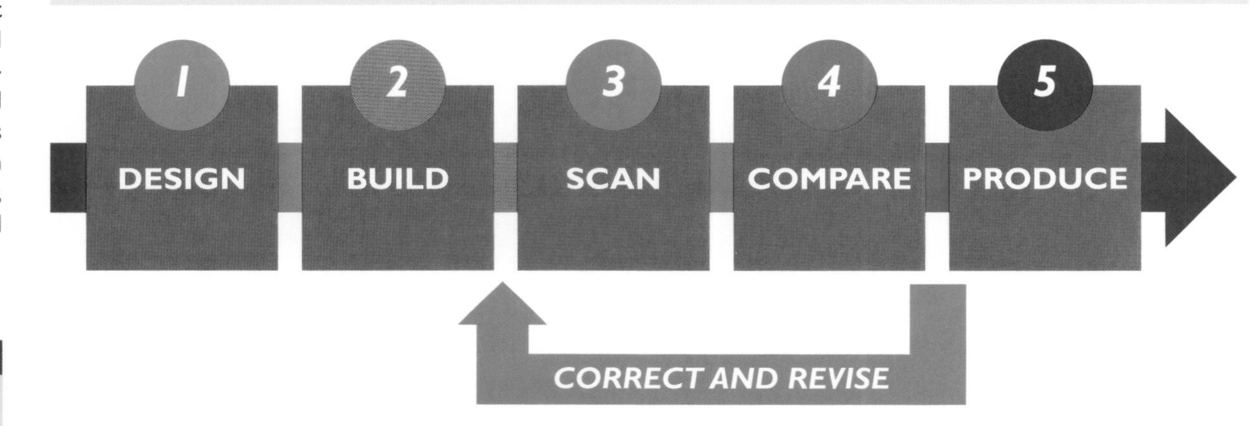

TOOLS, TACTICS & TALENT

- > *Develop virtual compliance analysis process*
- > *Integrate virtual with physical measurement*
- > *Use latest dimensional analysis tools*
- > *Coordinate and collaborate with suppliers*
- > *Develop specialists for implementation*
- > *Minimize use of design engineers for analysis*

RESULTS & BENEFITS

- > *Faster discrepancy identification*
- > *Optimized form and fit resolution*
- > *Improved manufacturing processes*
- > *Faster and better communication of results*
- > *Increased productivity of development engineers*
- > *Appropriate use of technology for better revenue*

Two Porsche beauties

NOTES:

Virtual Tolerance Analysis: Visual evaluation capability greatly improves and speeds the validation process

High-performance automobiles and motorcycles require crafted precision components for optimum quality and reliability. The castings in such vehicles must be carefully designed and manufactured to meet stringent dimensional specifications. Companies such as Volkswagen, Audi, Porsche, and BMW will accept nothing less than excellence in their vehicle components. The casting process for these parts at Rautenbach Aluminium-Technologie GmbH is sophisticated and precise. To validate the dimensional conformance of the resultant castings, a thorough process is used that combines both computer analysis and physical measurements. After casting and machining of the units, samples are digitally scanned and compared to the original virtual design computer models via software that identifies dimensional discrepancies. Precise physical measurements are also made to check certain characteristics. However, the great value of the digital comparison process is in having a comprehensive dimensional analysis of a part rather than only a few selected dimensions, and the quick visual evaluation capability via computer imagery.

Carsten Rudolf: Project Manager, Computed Tomography, Rautenbach Aluminium-Technologie GmbH

Carsten studied at the Fachhochschule Magdeburg (University of Applied Sciences) in Magdeburg, Germany in the field of mechanical engineering and design. Prior to working at Rautenbach, Carsten was employed at the Fraunhofer Institute of Factory Operation and Automation in product and process management. His research and development experience includes rapid prototyping, rapid tooling, CAD, reverse engineering, design, and measurement. Part of his current responsibilities at Rautenbach includes precision automotive casting tolerance evaluation and analysis.

Permanent mold casting carousel

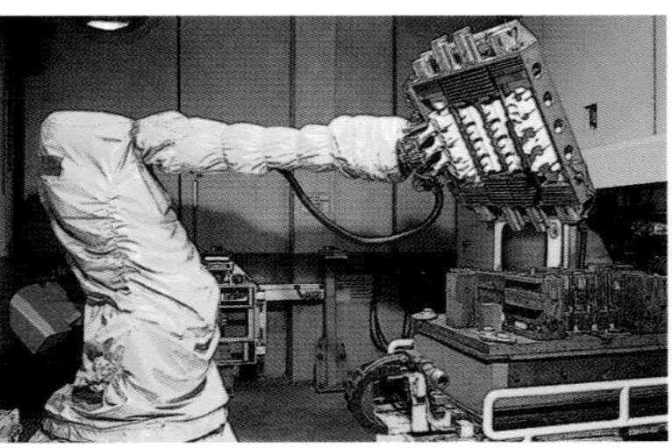

Core extractor robot

Leak testing device

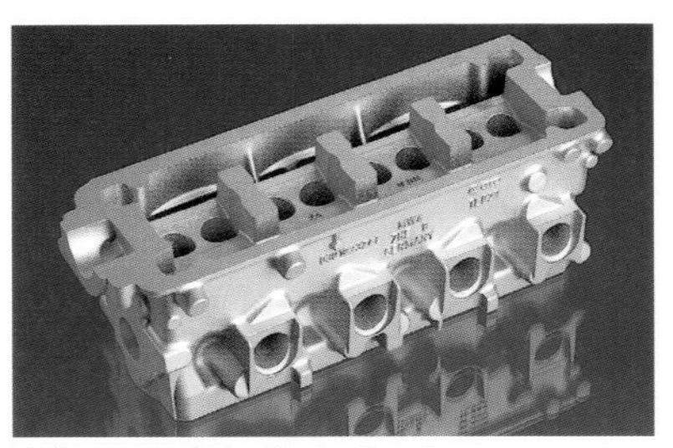

Volkswagen 2V cylinder head

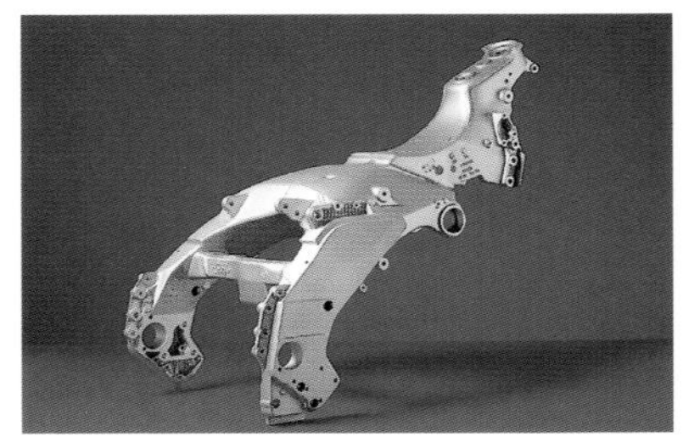

BMW motorbike frame

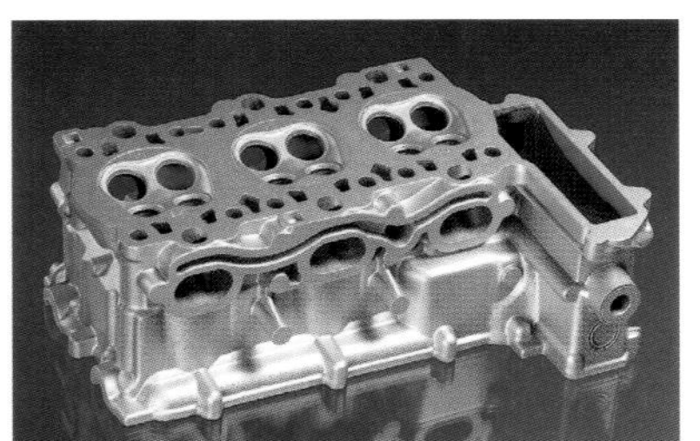

Porsche 4V cylinder head

RESULTS & BENEFITS

- > *Dramatic effect on speeding time-to-market*
- > *Reduced first-article-inspection (FAI) time from six weeks to one week*
- > *Savings on scrap by using non-destructive analysis*
- > *Reduction in prototype cost due to early virtual analysis and error correction*
- > *Expanded simulation applications using the precision analysis models*

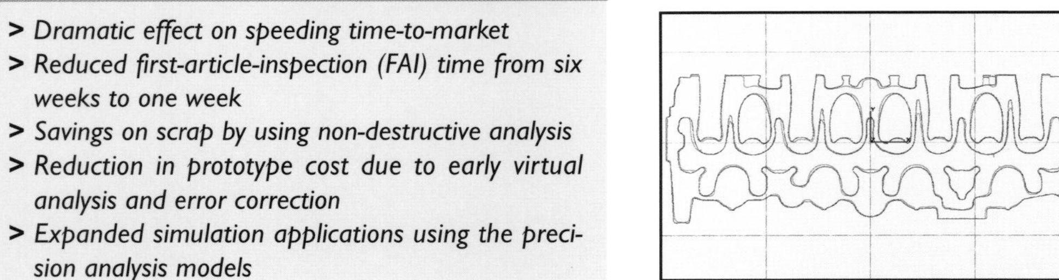

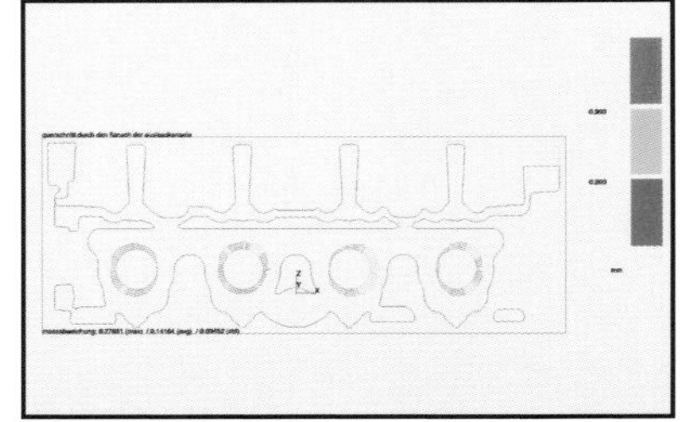

Two-dimensional tolerance analysis

Three-dimensional tolerance analysis

NOTES:

COLLABORATIVE EQUITY

> *"It's simple: you just take something and do something to it, and then do something else to it. Keep doing this, and pretty soon you've got something."*
>
> Jasper Johns

Entrepreneurs and design firms collaborating in product development and manufacturing

ROI usually refers to the return on investment a product developer makes in innovative design and implementation of a product concept. When the product is the first of an undercapitalized entrepreneur, that investment requires dedicating precious cash at a time when there may be little or none to invest. The ROI valuation is of only academic interest if no money is available. Traditionally, that's where a venture capital firm steps in. VCs prefer to maintain a high degree of control over the process, but are they equipped to manage the innovation process? There are those who are, but they are in the minority. Most are primarily finance and operations specialists. That's where Collaborative Equity comes in. A cash-strapped but idea-rich entrepreneur teams with an innovation-savvy, implementation-focused product design firm to actualize the product. Instead of handing equity (and control) over to a VC, the entrepreneur invests equity with the people who will generate the most innovation per dollar. The results can be startling and highly profitable for all concerned.

TOOLS, TACTICS & TALENT

- > *Undercapitalized entrepreneur*
- > *Alliance with innovative development specialist*
- > *Capitalize on entrepreneur drive and idea*
- > *Early involvement of development specialists*
- > *Early focus on innovation, not finance*

RESULTS & BENEFITS

- > *Earlier actualized, better, cheaper product*
- > *Maximized product innovation from all sides*
- > *Less managerial interference/control issues*
- > *Entrepreneur shares creative control*
- > *Equity and profit shared by actual developers*

Intrigo, Inc.: 350 Conejo Ridge Avenue, Thousand Oaks, California 91361

Founded in April 1998, Intrigo was conceived to become the leading brand of mobile computing workspace solutions. Privately held, Intrigo is based in Thousand Oaks, California. Partners include RKS Design, Inc., an award-winning product development consultancy, and Peerless Injection Molding, a high-technology product manufacturing firm. The primary co-founders and development leaders in the company are Maxim Weitzman, Intrigo CEO and director, and Ravi Sawhney, president of RKS Design.

TYPICAL VENTURE CAPITAL DRIVEN DEVELOPMENT SCENARIO

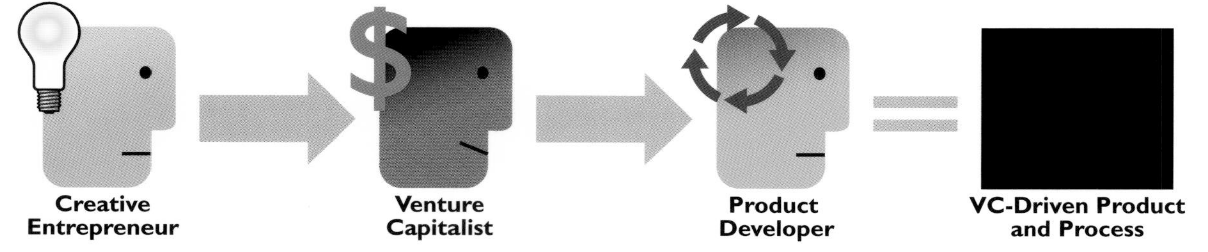

DESIGN DRIVEN COLLABORATIVE EQUITY DEVELOPMENT SCENARIO

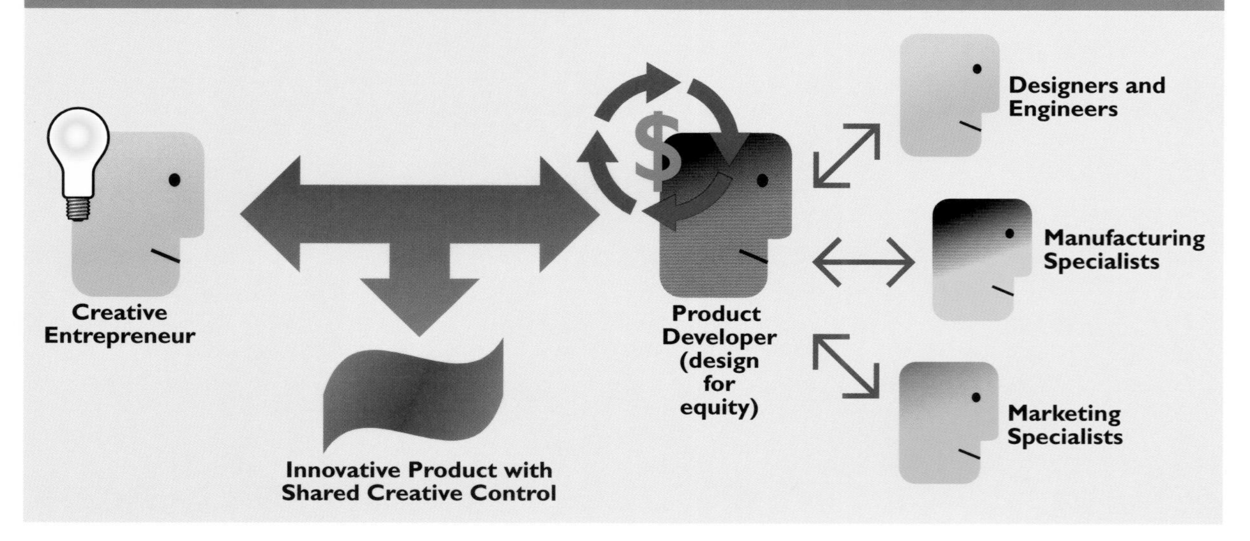

NOTES:

Joining Forces: The entrepreneur/designer collaborative venture for mutual profitability

Intrigo is an example of the synergy of Collaborative Equity. Fresh from his Entrepreneurial MBA program at UCLA, Maxim Weitzman had a great product idea and almost no capital. The solution? A strategic alliance with RKS Design, Inc. Ravi Sawhney, founder and president of RKS, took an equity position in Intrigo in lieu of cash for his firm's design services, producing an impressive design and prototype and introducing Weitzman to his ultimate tooling and manufacturing partner as well. The results? Quicker to market with a more innovative design and all realized below the market's expected price point. Here, the entrepreneur gets a better designed and implemented product with minimum impact on operating capital. The design firm has a more significant upside financial potential and more managerial control over the innovation process: a powerful combination of motivations to produce spectacular products.

Maxim Weitzman: Founder/CEO/Director, Intrigo, Inc., Thousand Oaks, California

Besides founding Intrigo, Inc., Maxim Weitzman has been responsible for developing and coordinating nationwide marketing strategies for several consumer product brands. While working as brand manager for Empresas Polar, Venezuela's largest privately held corporation, Maxim led the development of two products that became leading brands within their categories and won national advertising awards. Maxim holds a B.S. in industrial engineering and an M.B.A. from UCLA. In 1998, while at the Anderson Graduate School of Management, he served as director of programs of the Entrepreneur Association and received the Larry Wolfen Entrepreneurial Spirit Award for an independent start-up project. That same project represented the beginning of Intrigo. Maxim has been studying different trends within the computing workspace industry for the last three years and put this in-depth analysis to effective use in the development of the Intrigo Lapstation and its forthcoming accessories.

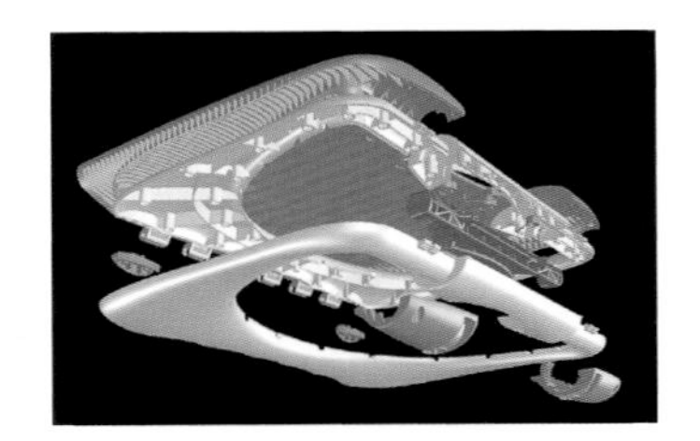

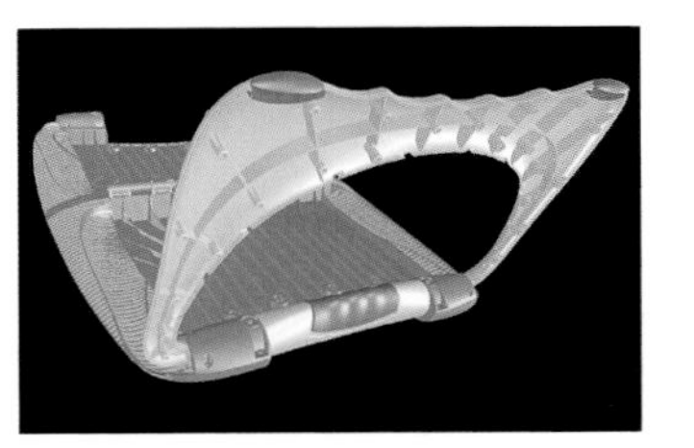

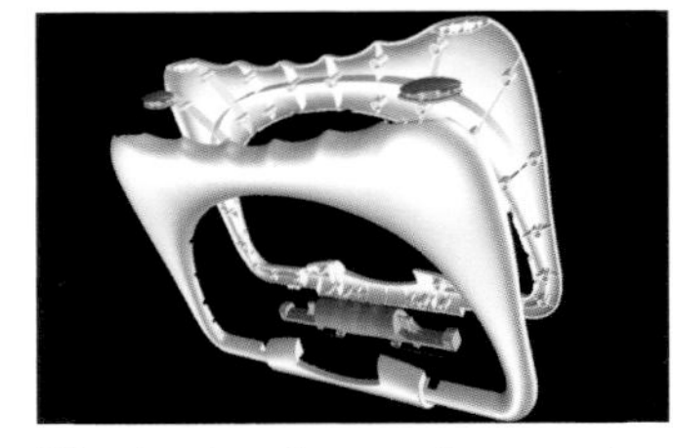

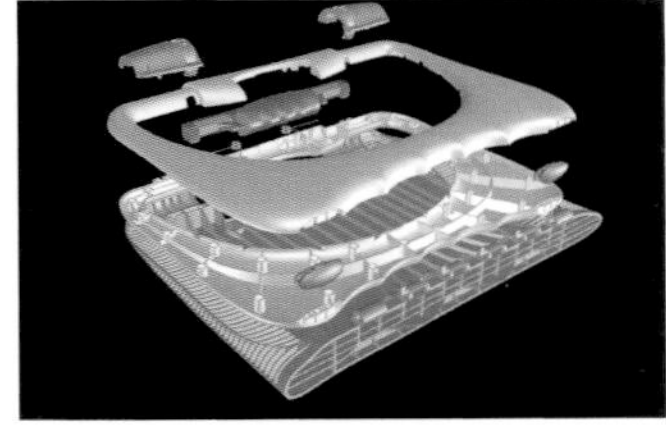

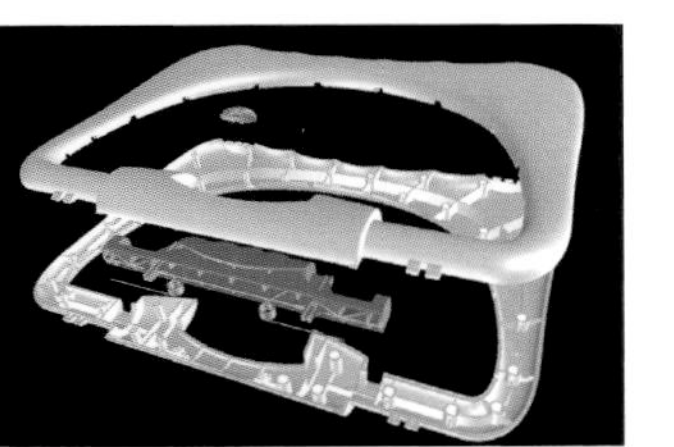

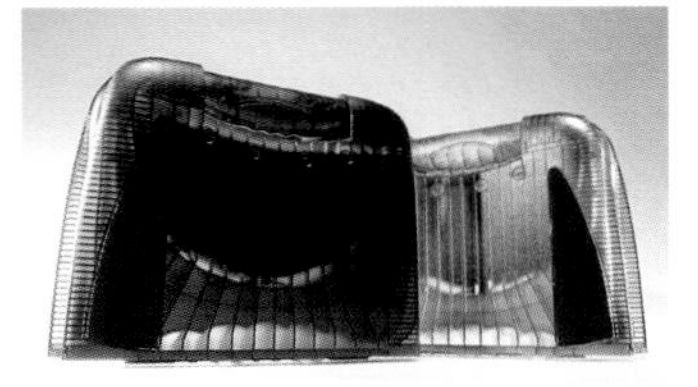

The Intrigo Lapstation product in various forms, parts, configurations and views

NOTES:

"I've seen corporate executives literally grow sick with fear at the prospect of having to propose and pursue a revolutionary new idea—which is, in the big picture, the only kind of idea worth having."

Doug Hall in *The Maverick Mindset*

"The business of business is ideas."

Jerry Hirshberg in *The Creative Priority*

"Great things are often done by naive people. Skeptics always know the cost of everything."

Henry Kissinger in television interview

BUSINESS & MARKETING

MOVING RESOURCES FORWARD

"When asked what single event was most helpful in developing the Theory of Relativity, Albert Einstein replied, 'Figuring out how to think about the problem'."

W. Edwards Deming

Adequate funding and resourcing early in a project to insure optimum performance

"The maximum potential for any design program is defined by the earliest phases those where multi-disciplined teams discover the greatest opportunities and identify the most far-reaching issues. Decisions from these phases define the functional platform and narrow path for refinement towards production. Any further changes become increasingly expensive propositions, as decisions build upon each other. If these early phases are underfunded, underresourced, or reduced in time allotment, the long-term costs can be significant. This can lead to missed opportunities (opening the door to competitors), higher costs, and even longer development time due to issues downstream that could have been addressed earlier with more time and expertise support. To maximize innovation and the return on investment in design it is imperative, even at the risk of overcompensation, to allow the time, resources, and funding for a complete design process as early as possible." Bob Del'Ve

TOOLS, TACTICS & TALENT

- > *Adequate loading of ideation and iteration*
- > *Heavy emphasis on simulation of ideas/concepts*
- > *Risk failure frequently and freely*
- > *Intense exploration, conceptualization, and debug*
- > *Engage as many participants/input as possible*
- > *All departments cross-functionally involved*

RESULTS & BENEFITS

- > *Early resolution of problems and issues*
- > *Development solutions at a low cost level*
- > *Early team-building and positive relationships*
- > *Comprehensive course for entire project*
- > *Optimized project performance at all levels: cost, schedule, innovation, quality, marketing, sales*

Robert Del'Ve: Senior Vice President, Designworks/USA (BMW), Newbury Park, California

Bob Del'Ve is a veteran sales and marketing executive who has helped business teams succeed in extremely competitive markets, integrated products and services through skillful planning, and developed and implemented profitable systems for several well-known companies. He now applies this expertise at Designworks/USA for its international product and transportation clients. Prior to his current position he was area sales director for Steelcase, Inc., and had the responsibility of accomplishing the corporate mission by focusing on customer service, quality control programs, and business development. Bob has an associate of applied science degree in architecture, a B.A. in business management, and an M.B.A. with a marketing emphasis.

A RULE-OF-THUMB RELATIVE COST EFFECT OF DESIGN CHANGES

During Concept	***Nil ("changes" are free)***
During Design	***1:1 (changes require the time to change)***
During Manufacturing	***1:10 (changes impact many systems in-process)***
After Shipping	***1:100 (changes impact many, many systems now out in market)***

THE IMPACT OF FUNDING AND RESOURCES EARLY OR LATE IN A PROJECT

The choice of front-loading for optimum results or panic back-loading when it's too late.

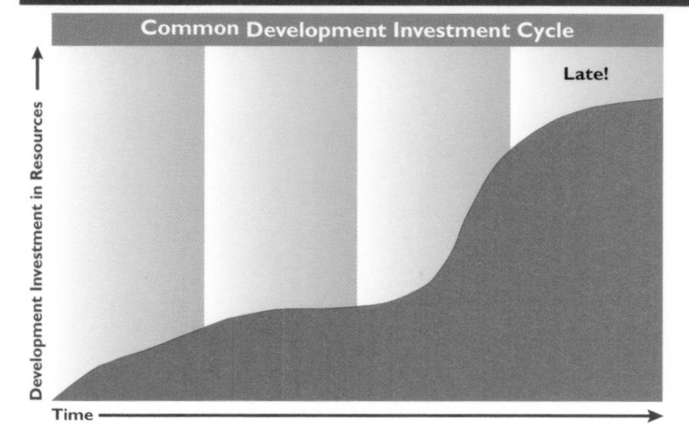

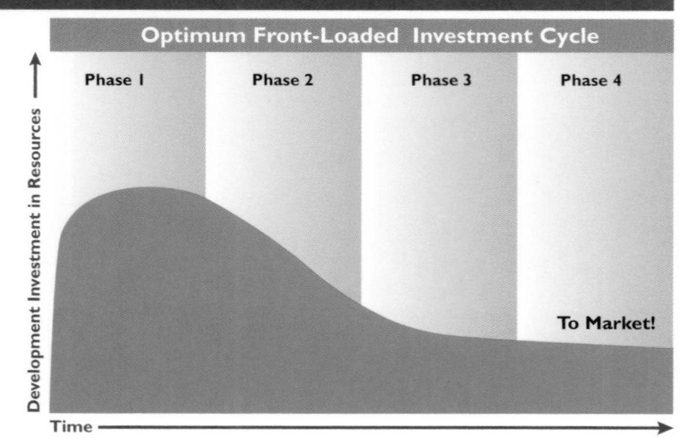

NOTES:

Forward Thinking: Development data that indicates where funding and resources are best invested

Development organizations focusing on lowest product cost, reduced development time, and competitive product features, will maximize ROI. This enables a company to achieve higher margins over a longer market window. Though seemingly obvious and inherently easy, most companies fail to maximize ROI because they attempt to serve all markets. Thus, fewer resources are available to develop core products, causing longer schedules. With inadequate early development resources, product optimization for low cost, high quality, and market-driven features takes longer. Rather than extend schedules, most companies sacrifice one or more of the above ROI maximization variables to ship product. Instead, they should focus on providing adequate resources for development while leveraging core competencies. To insure adequate funding for core projects, a company may need to cut lower-yielding, non-leveraged efforts ("zero-based budgeting"). Per empirical data, it is more important to strive for the lowest cost, highest price, and increased unit shipments versus development spending.

TOOLS, TACTICS & TALENT

> *Move resources forward into early development*
> *Fully fund and resource core product development*
> *Accelerate investment in tools, tactics, and talent*
> *Leverage core competencies*
> *Emphasize zero-based budgeting*

RESULTS & BENEFITS

> *Reduced project costs and schedules*
> *More innovative and feature-laden products*
> *Reduced time-to-market*
> *Increased revenue from earlier market entry*
> *Improved ROI*

Brian Marx: Finance Manager, InFocus Corporation, Wilsonville, Oregon

Brian Marx is always looking for innovative ways for a company to improve its bottom line and he has been able to pursue this endeavor as a finance manager for InFocus. There, he is responsible for maximizing product line profitability and return on investment and has successfully designed and implemented the corporate Project Accounting System. Brian has previously worked for National Semiconductor Corporation as a product line controller and Westinghouse Electric Company as a cost analyst. Brian received a B.S. in finance and also has an M.B.A.

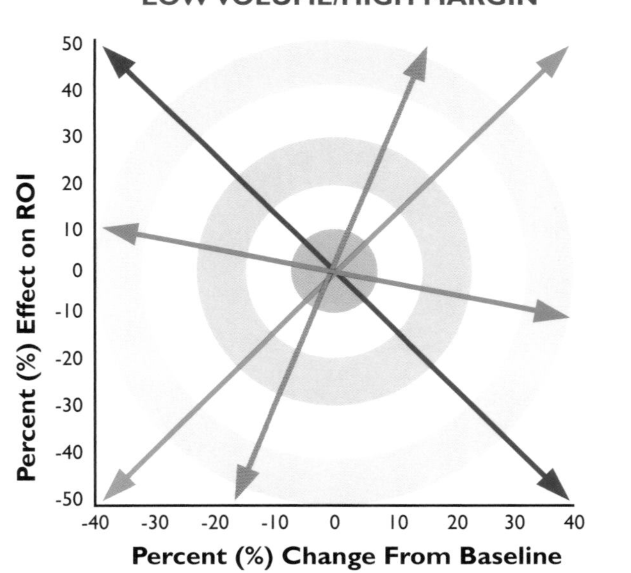

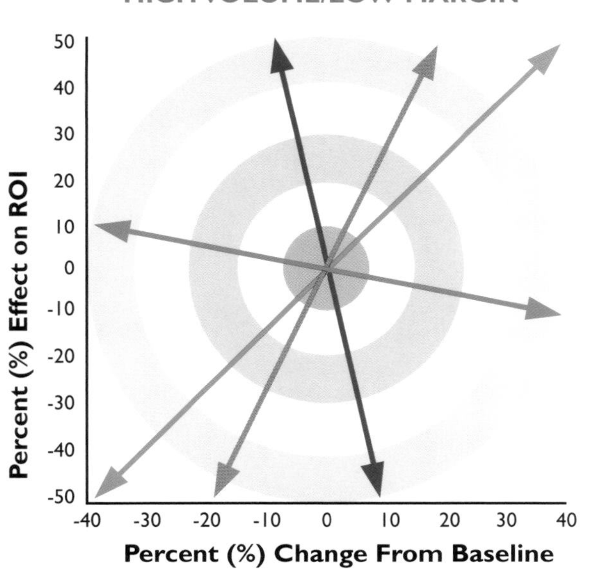

NOTES:

"The past has been a horribly misleading guide to the future."
Tanya Beder

DEVELOPMENT INVESTMENT ROI

Properly understanding return on investment for resources based on increased product revenue

If you're responsible for a project budget, you know the nagging fear of being second-guessed on each expenditure. Spending a budget means justifying every dollar as necessary to the desired outcome, and money spent on innovation is no exception. So how do you justify spending money on the processes described in this book? Your investment in tools, tactics, and talent must be defended based on outcomes like reduced time-to-market, first-mover advantage, higher perceived product quality and functionality, and product differentiation. Each of these outcomes will ring the cash register with traceable, measurable revenues that justify your initial expenditures. Then why is it so difficult to get approvals for these activities? The graphics on this page tell the story. Most managers won't understand the process until you "open up the black box" and show them how investments in innovation add directly to the bottom line.

TOOLS, TACTICS & TALENT

> *Implement best practices development policy*
> *Invest in state-of-the-art development tools*
> *Attract, hire, and keep world-class performers*
> *Measure ROI by revenue output vs. resource investment input, not initial cost of resources alone*
> *Apply appropriate resources early and aggressively*

RESULTS & BENEFITS

> *Useful and realistic ROI on resources*
> *Improved overall development process*
> *Enterprise-wide rationale for resource allocation*
> *Management support for best tools and talent*
> *Maximum bottom line revenue performance*
> *Enterprise understanding of true resource value*

COMMONLY MISUNDERSTOOD DEVELOPMENT MODEL

The Problem:

ROI = ?

To many business and finance people, the product design, development, and innovation process is a black box, poorly understood at best. Consequently, computing an ROI for development investment is difficult.

COMMONLY UNDERSTOOD INVESTMENT MODELS

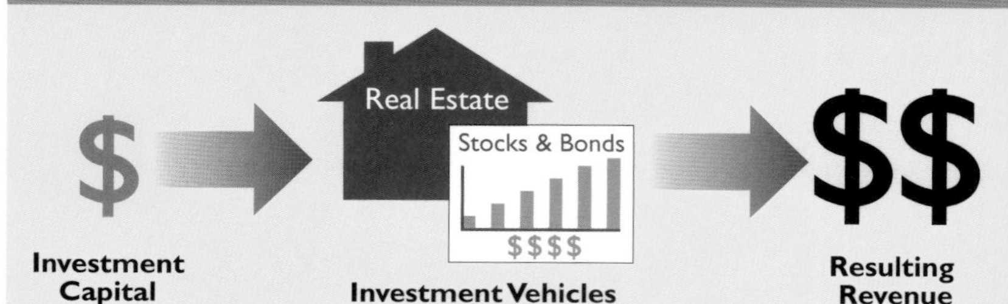

ROI = \$\$\$ / \$ = !

However, most investors understand the opportunities and ROI of other types of investments as well as the process. Here, ROI is simply the ratio of output value divided by input value.

PRODUCT DEVELOPMENT INVESTMENT MODEL

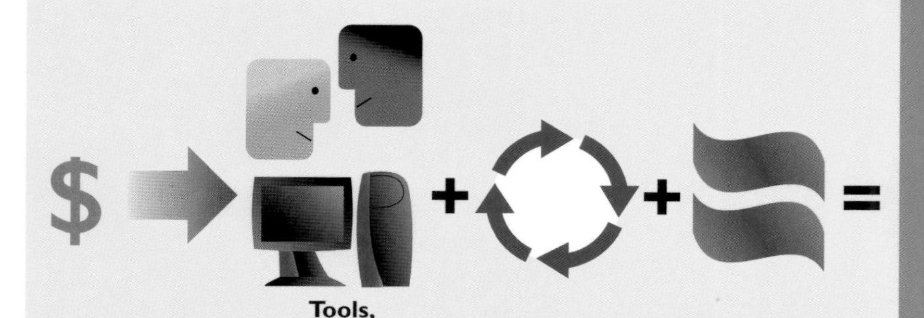

ROI = \$\$\$ / \$ = !

The ROI for product design and development and the investment in tools, tactics, and talent for innovation is no different. It's product revenue output value divided by input investment value.

\$\$\$

Resulting Revenue

NOTES:

Resources ROI: Proper understanding and evaluation of resource costs and benefits are critical to success

Within many corporations funding design and development resources, whether internal or external, can be the single largest budget line item that management knows the least about. So it's easy to understand why they become suspicious about the payoff on their investment. A common philosophy is to avoid going outside unless absolutely necessary (i.e., a crash-and-burn, too-late, emergency situation!). Retaining consultants is often seen as inefficient, costly, risky, and/or too time-consuming. This does not have to be the case, especially if internal resource costs are well understood. Properly integrated, consultants can offer fresh perspectives, flexibility, and a diversity of experience that might not be available inside the company. Capitalizing on external expertise can be very effective if the resource deployment process is managed correctly. Every company should have someone—with practical design experience—to direct and protect its ROI. To that end, an astute internal design manager delivers as much value as any good financial manager.

TOOLS, TACTICS & TALENT

> *Consideration of creativity/innovation factor*
> *Evaluate expertise available in-house*
> *Consideration of internal bureaucracy factor*
> *Evaluation of distance factor for outside services*
> *Analyze true cost and overhead factors*

RESULTS & BENEFITS

> *Reduced overall development costs*
> *Increased speed of completion*
> *Optimally blended inside/outside teams*
> *Mutual reinforcement of strengths*
> *Weaknesses minimized by collaboration*
> *Increased knowledge/information base*

Maureen Thurston: President/Founder, ACCESS International, Inc., Pasadena, California

Ms. Thurston's personal and professional goal is to promote the value of design to industry, reinforcing its competitive contribution to the bottom line. She received her B.F.A. in industrial design from Rochester Institute of Technology and is president and founder of ACCESS International. Established in 1987, ACCESS was the first brokerage for product design and development in the United States, and has since evolved into a design strategy consultancy helping manufacturers integrate design methodologies into their competitive business strategies. Author, guest speaker, and instructor of design management at Art Center College of Design, she is also managing director of Practical Products, a design/licensing company specializing in houseware products.

INVESTMENT FORMULA

	timeline	hours/month	x 80% actual	x # specialists	= total hours	x charge-out rate	= projected low	= projected high
Straightforward (e.g., desktop accessory)	2-4 months	320-640 hours	256-512 total hours	Two people	512-1,024 person-hours	$80 - $120 per hour	$40,960 - $61,440 (2 people x 2 months)	$81,920 - $122,880 (2 people x 4 months)
Midrange (e.g., consumer electronics)	5-8 months	800-1,280 hours	640-1,024 total hours	Four people	2,560-4,096 person-hours	$80 - $120 per hour	$204,800 - $307,200 (4 people x 5 months)	$327,680 - $491,520 (4 people x 8 months)
Complex (e.g., medical instrumentation)	9-12 months	1,440-1,920 hours	1,152-1,536 total hours	Eight people	9,216-12,288 person-hours	$80 - $120 per hour	$737,280 - $1,105,920 (8 people x 9 months)	$983,040 - $1,474,560 (8 people x 12 months)

BOTTOM LINE: The ROI should be based on the increased product revenue resulting from the use of adequate and appropriate development resources.

NOTES:

Tools ROI: Choosing the right tools for the job involves return on investment evaluation

Since computers were developed as tools, product development time has dropped from two years or more to nine months or less. But it isn't the computers, or the software, or the new 3D printers that are, by themselves, making the difference. These specialized, sophisticated high-speed wonders require a change in thought process and methodology, and it is that change which has triggered a whole new way of developing products. The tools are important—indeed, without them much of what has been accomplished would have been impossible. But it's the total process that makes the difference. What hardware and software should you buy to make these huge capability and functionality leaps? Whatever brand it turns out to be, plan on buying the best. The initial dollars saved with a discount system approach that's 'good enough' will pale by comparison to lost time-to-market, worker frustration, and ultimate system replacement.

TOOLS, TACTICS & TALENT

> *Best-in-class appropriate tools*
> *Best practices process and methodologies*
> *Intense continued training and maintenance*
> *World-class professional talent*
> *Adequate support base and personnel*
> *Total system integration of tools and tactics*

RESULTS & BENEFITS

> *Highly productive and motivated staff*
> *Improved professional recruiting advantage*
> *Improved overall development process*
> *Design tools as strategic business advantage*
> *Improved time-to-market*
> *Increased revenue from faster TTM*

Tom McKasson: Founder and President, Acuity Corporation, Portland, Oregon

ACUITY
ACUITY INCORPORATED

In the late 1980s, Tom McKasson applied CAD, CAM, and manufacturing data management to the newly emerging market of inexpensive, personal computers. Tom and his wife, Anne, then implemented CADKEY, a PC-based 3D mechanical design software tool, into the Oregon design market and established their company, Acuity Corporation. Tom continues to help businesses best utilize the most recent technologies. Acuity has become one of the most successful value-added resellers in the product development and data management market. Acuity's partners include such companies as Alias|Wavefront, CADKEY, Intergraph, Silicon Graphics, 3D Systems, and SDRC. Acuity helps its customers benefit from the newest technologies in product design and development by using "digital clay," 3D printers, CAID and CAE software, and other innovation tools.

TWO TIME LINES

DEVELOPMENT WITH LOWER INITIAL COST COMPROMISED TOOLS:

Buy Tools | Learn Tools | Less Effective Productivity with Tools in Product Development | Results

Time-to-Market (TTM) | Revenue

Additional Costs

DEVELOPMENT WITH SUPERIOR AND APPROPRIATE TOOLS:

Buy Tools | Learn Tools | Productivity with Tools in Product Development | Results

Time-to-Market (TTM) | Revenue

Additional Costs | TTM Advantage

BOTTOM LINE: Can you spend initial extra tens of thousands of dollars on tools to make additional millions of dollars due to the TTM revenue advantage?

NOTES:

Balanced Life ROI: Achieving personal and career balance for optimum productivity and innovation

Managers have often wondered why some people are more innovative than others. Glen and Liz Kauk believe it has a great deal to do with balancing what they call The Wheel of Life™. The Kauks, principals of Bright Future Business Consultants, consult with businesses about improving their return on investment by improving employee productivity. A particular focus of their work is helping employees create more innovative products or services in large corporate and government environments where a high level of stress is prevalent. The key to being more innovative in such environments, the Kauks propose, is having life in balance. Addressing all of life's goals, not just job or career goals, is central to freeing the mind from the constraints and pressures of an intense job-only mentality. Balanced employees, whose personal goals mesh with career goals, are more likely to generate new ideas in all areas of life, including their work.

TACTICS, TOOLS & TALENT

- > *Consciously commit to time for personal life*
- > *Deliberately manage non-work time*
- > *Dedicate time for relaxation/contemplation*
- > *Develop a clear priorities list and follow it*
- > *Incorporate spiritual aspects into daily life*
- > *As much as possible, do what you love*

RESULTS & BENEFITS

- > *Increased productivity and innovation*
- > *More time for creative activities*
- > *Personal contentment and satisfaction*
- > *Higher motivation level*
- > *Improved growth and development*
- > *Better relationships and job satisfaction*

Glen and Elizabeth Kauk: Principals, Bright Future Business Consultants, Union City, California

Glen

Liz

For the past six years, Glen has been vice president of Bright Future Business Consultants, which provides dedicated personal and professional development services. Clients include high-technology Silicon Valley companies, municipalities, factories, and religious organizations. Prior to BFBC, Glen worked for twenty years in the retail grocery business. When Glen is not assisting clients, he enjoys restoring classic cars.

Liz is vice president and CFO of Bright Future Business Consultants. She has taught college business courses, drafted and negotiated world-wide distribution contracts for a high-technology multi-media company, administered a venture capital fund, and co-founded and managed a travel agency. Liz has an A.S. in landscape horticulture, a B.S. and an M.S. in mathematics and computer science education, and an M.B.A. in finance.

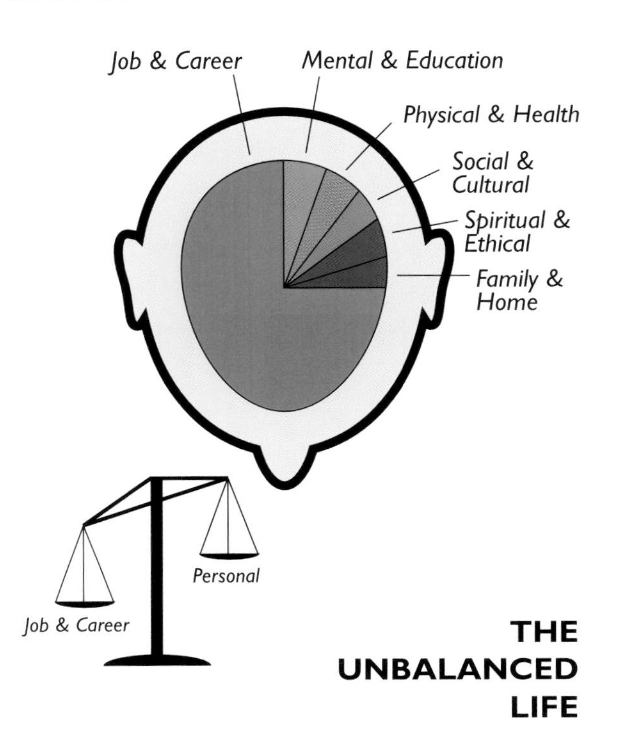

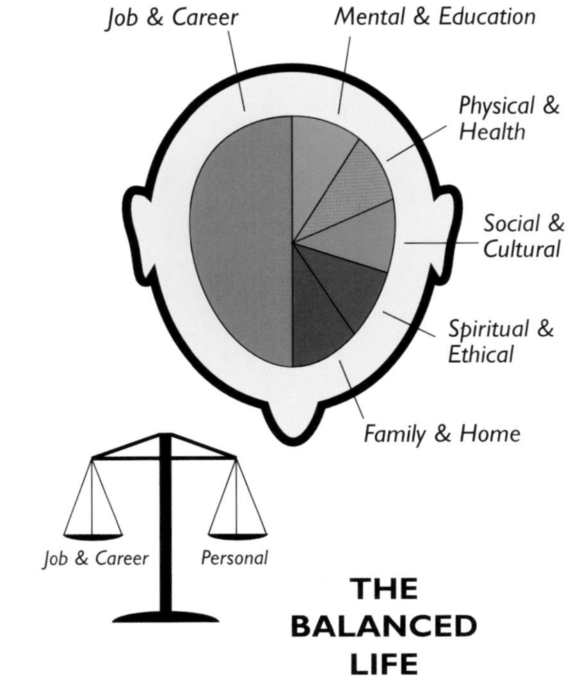

NOTES:

Today's competitive marketplace demands strategic product design as a key factor in success

Ask most people (especially engineers, finance specialists, and other "hard science" types) what product design is, and they'll likely respond with something about making the product look good. Ask a designer the same question and you'll get a more comprehensive answer. Effective product design develops great looking products because it first concerns itself with great functioning products that have innate appeal and value to their users. Beauty is more than skin-deep. Product design is a creative way of resolving challenges faced by users of products. Consequently, product design organizations bring resources to business that are valuable in a wide range of applications. Indeed, great design is a corporate strategy for companies that want to grow by developing long-term relationships with a large body of satisfied users.

TOOLS, TACTICS & TALENT

> *Establish design as a company core competency*
> *Approach design as a critical corporate strategy*
> *Invest in best design tools, tactics, and talent*
> *Create a corporate culture of design*
> *Use design as a business tool for the enterprise*
> *Eliminate design as a product cosmetics tool*

RESULTS & BENEFITS

> *Quality design across the enterprise*
> *Corporate bottom line and long-term impact*
> *Insight into user/product relationship*
> *Defined brand identities for market advantage*
> *Design used as powerful problem-solving tool*
> *Business-driven design and design-driven business*

Sohrab Vossoughi: Founder/CEO, Ziba Design, Inc., Portland, Oregon

Sohrab graduated from San Jose State University in 1979 with a B.S. in industrial design and is experienced in the field of mechanical engineering. After graduation, he began working for Hewlett-Packard and subsequently became an independent contractor in 1982. In 1984, he established ZIBA Design, which holds over thirty patents and 200 design awards. Sohrab was distinguished as BusinessWeek's 1992 Entrepreneur of the Year, recognized as one of the forty most influential designers in the United States by International Design Magazine in 1994, and is the only industrial designer ever elected "Global Leader of Tomorrow" by the World Economic Forum.

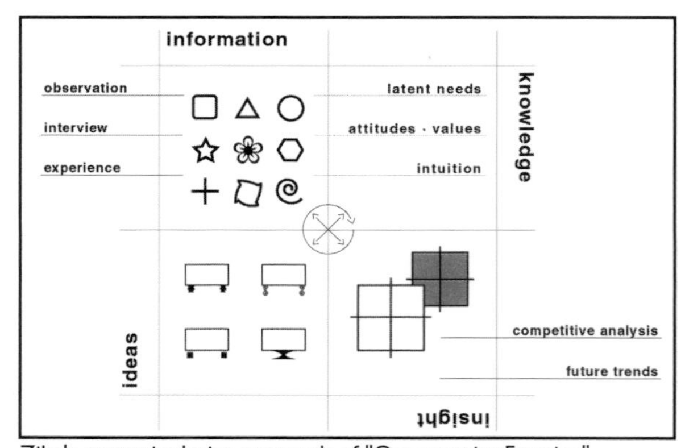

Ziba's strategic design approach of "Opportunity Framing"

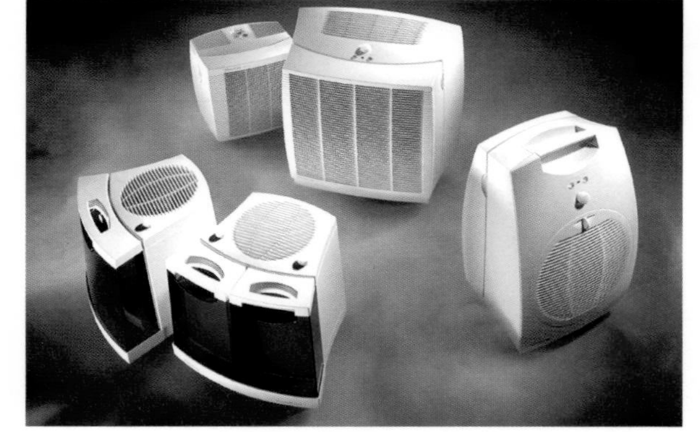

Strategically designed Rival Bionaire Air Quality System product line of air purifiers and humidifiers

NOTES:

Design Research: A powerful corporate strategy tool applied to product development

Design. Research. These go together? The first seems creative, the second academic. Not so. Katherine Bennett knows: "Product designers live at the intersection of culture and commerce. We make it our business to know what makes a product truly useful, and beyond that, what makes it desirable. The research we do enables us to understand the problem clearly. It is the foundation of good design. New methodologies, like videoethnography, we've borrowed from anthropologists, and others, like building mockups and testing them with consumers, we have been using for years. Traditional market research techniques aren't enough any more. Using sterile focus groups and phone surveys guarantees getting the same tired answers as the competition. Industrial designers' research techniques get product developers out of the office and into the environment of product use, where we can see what is really going on."

TOOLS, TACTICS & TALENT

- > *Flexible and definitive research process*
- > *Application of research throughout a project*
- > *Adequate research tools, processes, and resources*
- > *Environments for research and interaction*
- > *Trained research specialists*
- > *Continual communication of research results*

RESULTS & BENEFITS

- > *Well-informed development staff*
- > *Efficient product development process*
- > *Competitive edge in marketplace*
- > *Increased product innovation*
- > *Precise market attributes identified*
- > *Direct market hits for product features*

Katherine Bennett: Principal, Katherine Bennett Industrial Design, Santa Monica, California

Katherine is a consultant and educator in product design and development. She provides design, research, and strategic planning for clients worldwide from her office in Santa Monica, California. Katherine is an active member of the Los Angeles chapter of IDSA and served as its chair from 1997 to 1999. She is also a faculty member of the product design department at Art Center College of Design and has lectured at several prominent schools including UCLA, USC, and the DMI educators' conference at Stanford University. Katherine received her B.S. in industrial design from the Philadelphia College of Art and attended the Pennsylvania Academy of the Fine Arts.

On-site investigation for camping equipment

Branding imagery study

Research data presentation and discussion

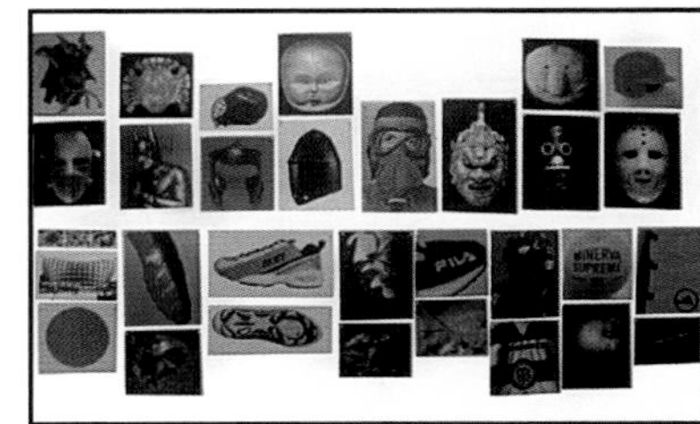

Ethnographic research image board

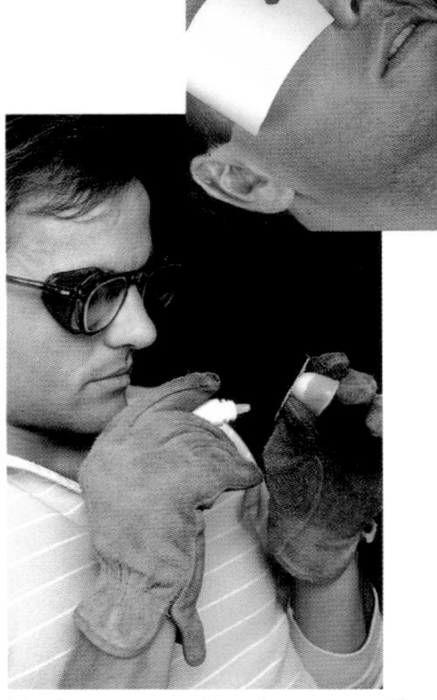

Testing ergonomics of eye drop dispensing

NOTES:

MARKET REVOLUTION

"A successful . . . demo isn't show-and-tell; it's show-and-ask."
Michael Schrage

Creating compelling product appeal and market dominance with revolutionary design

Sportswatches have existed for decades and remained relatively untouched by recent technology until Astro Studios and Nike teamed up to reinvent this flat product category. Product issues such as user-focused ergonomic design, product styling, interactive features, and unique configurations can make the difference between being a market leader or having just another me-too sportswatch. Astro's approach was simple: rethink the idea of sport timing from the ground up, break all the rules, and innovate at every level. The result has redefined the sports-specific timepiece and helped create a market revolution. This was done by inventing the Triax's "ergonomic S shape" that fits the natural contours of the wrist, permitting increased size and functionality of the display, and addressing button navigation with a "grip zone" that acts as a reference point for the user. Astro's successful iconic designs for Nike have revolutionized the sportswatch market.

TOOLS, TACTICS & TALENT

> *Early user research, interviews, and observations*
> *Extensive ideation concept sketches*
> *Quick 3D sketch models to explore ergonomics*
> *Focus groups for interaction and navigation tests*
> *3D CAID modeling of design surfaces and details*
> *Strategic alignment with manufacturing partners*
> *Empowered entrepreneurial start-up group*

RESULTS & BENEFITS

> *Best client sales with first-ever electronic product*
> *Design brings broad acceptance in target market*
> *Cross-over effect creates broader mass demand*
> *Imitation by competition due to extensive impact*
> *Design excellence awards from industry media*
> *Revolution in market impact and client revenue*

Astro Studios, Inc.: 818 Emerson Street, Palo Alto, California 94301

ASTRO

Astro is an experienced and talented group of industrial and graphic designers with expertise in consumer and computer lifestyle products. They develop products and identities for companies who seek the cutting edge of design and business. Astro was established in 1994 by Brett Lovelady, who was previously vice president of design and marketing at Lunar Design, design manager and vice president at frogdesign, and a senior designer with Tandem Computers. He was joined by Kyle Swen and Rob Bruce. Astro provides services, including industrial design, graphic/web design, engineering development, modeling and prototyping, and advertising and brand identity design, to such clients as Apple Computer, Compaq Computer, InFocus, Kensington, Polaroid, Motorola, and Nike.

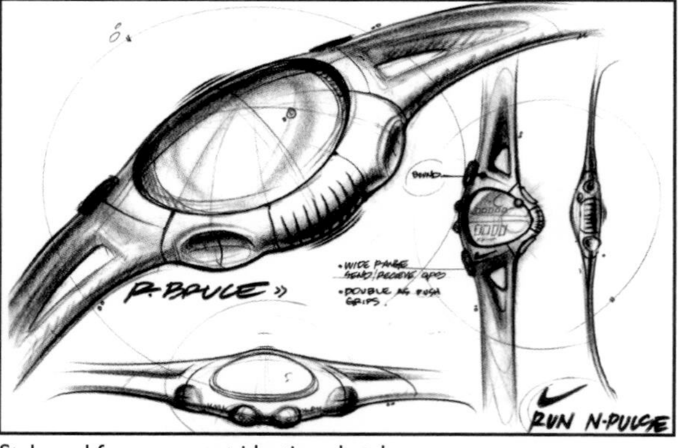

Style and form concept ideation sketches

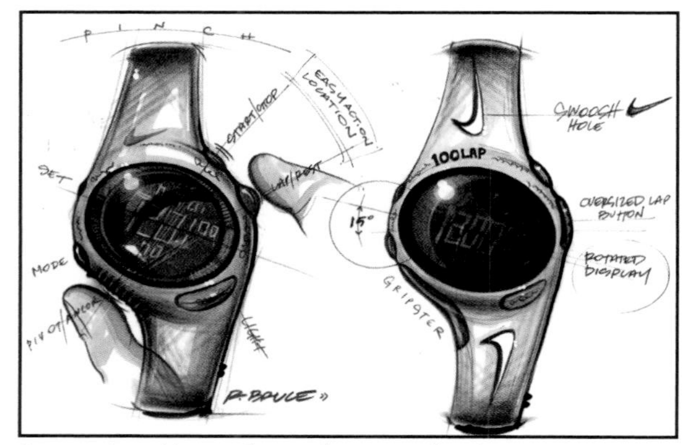

Ergonomic and human factors feature exploration

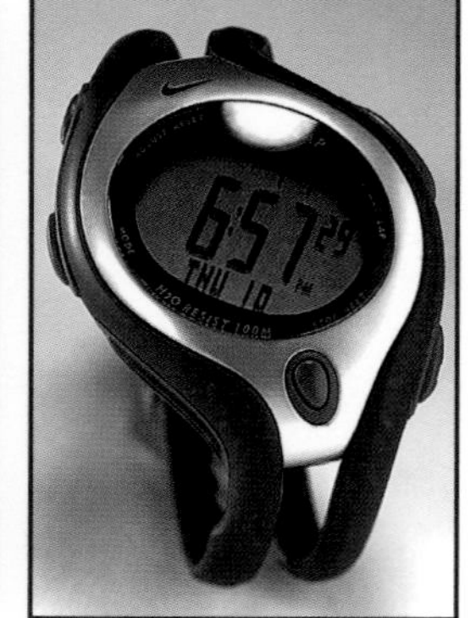

Multiple design variations with varied form, interface, features, and details

Subtle, crafted detailing

NOTES: ______________________

Psycho-Aesthetics™: The psychology of great product development and compelling product experience

Something happens psychologically to people who use well-designed products. Using such products makes people feel good about themselves as well as think positively about the product's source organization. RKS Design has coined a term for this syndrome: Psycho-Aesthetics™. It is, in their words, "the explicit promise of a product to deliver the intended value." This is, however, much more than simply exceeding the performance expectations of the customer. Does using your product make users happy? Do they feel good about themselves because they own and use it? Is the experience of using the product a rewarding and positive one? These questions are critically important to ask and rather difficult to answer objectively. RKS has a proprietary methodology for doing so, with the answers feeding an iterative development process that moves constantly toward providing users with psychological as well as economic value in the products they design.

TOOLS, TACTICS & TALENT

- > *Drive design from customer wants and needs*
- > *Understand and interpret customer input*
- > *Focus development on human needs*
- > *Constant iteration of process*

RESULTS & BENEFITS

- > *Customer-focused product designs*
- > *Improved marketability of products*
- > *Enhanced customer satisfaction*
- > *Customer and human need-driven design staff*
- > *Product innovation advantage*
- > *Optimized user experience design*

Chip Wood: Vice President, Strategic Design Planning, RKS Design, Inc., Thousand Oaks, California

Chip has twenty-seven years as a veteran industrial designer and design manager for three Fortune 100 companies and several design consultancies. Prior to joining RKS Design, he was the world-wide manager for design at Texas Instruments. There he was chosen to head up a new business research, planning, and design lab. Chip now assists RKS clients in identifying and building new business opportunities by using strategic design planning methods, including user research and analysis, business concept and scenario planning, and experiential business/product prototyping. The Psycho-Aesthetics™ process is part of this comprehensive support.

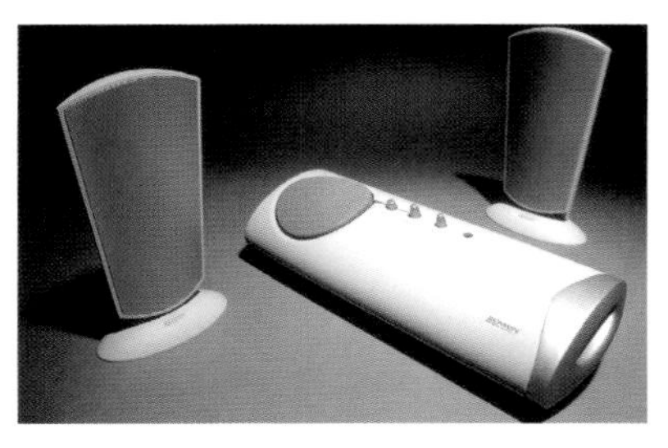

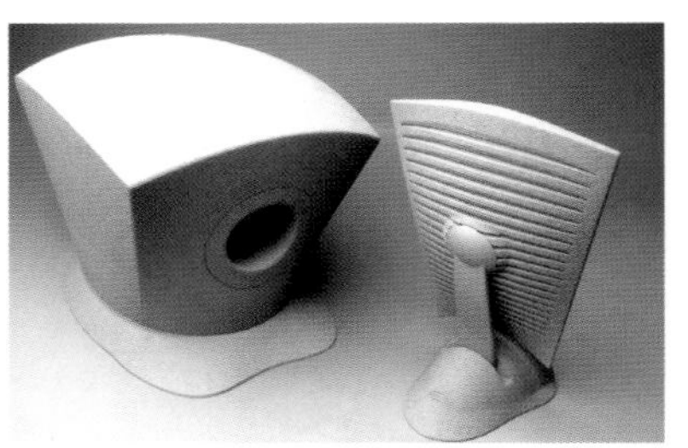

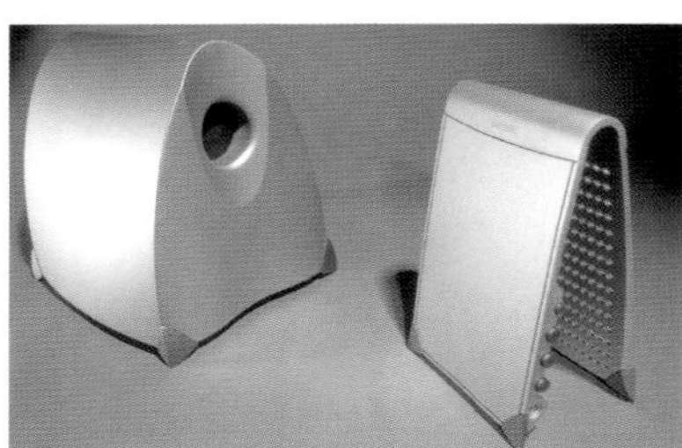

Several product images reflecting the application of Psycho-Aesthetics™ to product feature design and detailing

NOTES:

> "Persistence is the hard work that you do after you are tired of doing the hard work you already did."
>
> Newt Gingrich

DECISION TRIAD

Choices regarding quality/cost/schedule (QCS) for optimum development performance

Any company engaging in high-technology product development must constantly balance the value of product quality and innovation, development cost and investment, and project schedule and time-to-market. This balance must be constantly evaluated from early planning stages through product development to manufacturing and sales. Choices must be made as to which of these elements must be optimized. Specific decision sets will dictate certain outcomes. The simplified charts on this page indicate the consequences of picking two of three elements. The last chart shows the most common scenario for the aggressive high-technology market: world-class quality with fast time-to-market. Here, by spending appropriately to insure optimum product development with lowest market risk, the return on investment in profitability pays off. The first two charts show what happens to schedule and/or quality when development investment is compromised.

TOOLS, TACTICS & TALENT

> *Clear understanding of ROI objectives*
> *Agreement of quality, cost, and schedule definitions*
> *Understanding of QCS requirements/ramifications*
> *ROI evaluation of QCS options*
> *Appropriate budget planning around QCS choices*

RESULTS & BENEFITS

> *Optimized return on investment*
> *Best overall project performance*
> *Appropriate market positioning*
> *Proper product introduction timing*
> *Optimized product features and detailing*
> *Maximized profitability*

John V. Harker: President/CEO/Chairman, InFocus Corporation, Wilsonville, Oregon

John Harker has been at the helm of InFocus since 1992 and board chairman since 1994. He has held top level executive management positions at such companies as Genicom, Data Products, Booz-Allen, and IBM and has a Bachelor of Science in Marketing. As the head of the company that is the worldwide leader in data/video projection systems and services, John knows the tension of balancing development investment cost against return on investment for his shareholders. He has learned from long and successful experience that spending what it takes to get world-class product design and performance has a high return on investment value.

LOW COST, FAST PROJECTS CAN RESULT IN LOW PRODUCT QUALITY CONTENT

	Chosen	LOW — HIGH
(Product) Quality	☐	Low
(Development) Cost	☑	Low
(Project) Schedule	☑	Low

HIGH QUALITY PRODUCTS AT LOW PROJECT COST USUALLY TAKE LOTS OF TIME

	Chosen	LOW — HIGH
(Product) Quality	☑	High
(Development) Cost	☑	Low
(Project) Schedule	☐	High

HIGH QUALITY PRODUCTS EXECUTED QUICKLY REQUIRE ADEQUATE FUNDING

	Chosen	LOW — HIGH
(Product) Quality	☑	High
(Development) Cost	☐	High
(Project) Schedule	☑	Low

NOTES: ______________________

Choosing the Right Two: Decisions and funding for product quality and speed of development

Smart companies that must hit their market window as quickly as possible with the best products are aware it takes investment in tools, tactics and talent to achieve that goal. This does not mean they throw money away or throw money at a project. What it does mean is they carefully allot their project investment at the appropriate times and amounts to achieve optimum quality and schedule performance. This is done so their return on a high development investment yields a high profitability due to their quality and time-to-market advantage achieved. InFocus is one of these companies. They are the world's leader in data/video projection systems in a market that is incredibly aggressive, with over one hundred competitors worldwide. They have achieved this leadership by investing heavily and appropriately in product quality and innovative design and engineering to achieve killer products and aggressive schedules. Their most recent entry into the market, the LP330, has dominated the market as the lightest, brightest, and most attractive projector in the field.

TOOLS, TACTICS & TALENT

- > *Commitment to quality design execution*
- > *Adequate and early funding and resource loading*
- > *Emphasis on innovation and problem-solving*
- > *Use of state-of-the-art technology and tools*
- > *Commitment to impeccable productization*

RESULTS & BENEFITS

- > *Optimized project performance and speed for ROI*
- > *Best-in-class productization of technologies*
- > *Market leading product performance and brand*
- > *Invigorated and motivated development staff*
- > *Long term profitability and product leadership*

Phillip Salvatori: Industrial Design Manager, InFocus Corporation, Wilsonville, Oregon

Phil is industrial design manager at InFocus and is responsible for the look and feel of all IFS products. He manages both in-house and external industrial design resources for all projects. Prior to InFocus, Phil designed many products as a consultant including consumer, business, medical, and recreational products and has received several international design awards. Phil received his B.S. in industrial design from California State University at Long Beach and has a passion for undertaking seemingly overwhelming objectives and accomplishing them through product design innovation.

Compelling appearance and styling design

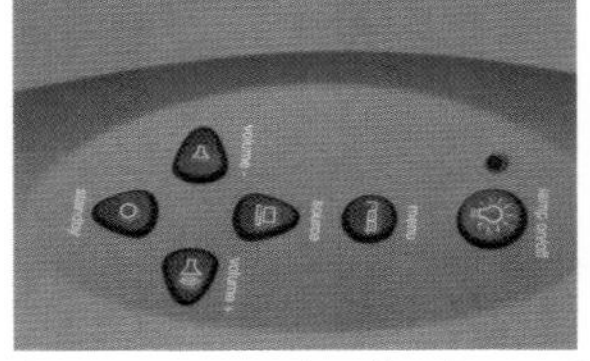

Attention to quality detailing and product language

Market brand and advertising leadership

Extremely compact custom product engineering

NOTES:

SIMULATION DRIVEN MARKETING

Successful early market research utilizing visualization and simulation tools and tactics

Focus groups and user interactions have been commonly used techniques for product market research for some time. However, the engagement process is often limited to primarily verbal interaction. Queries concerning user expectations and responses to existing products are the usual fare. Unfortunately, verbally-based interaction frequently does not lead to the kind of data that is required for useful product and feature definition. Such data often creates confusion and disparate results. With the many new digital computer simulation and rapid prototyping tools available today it is possible to realistically simulate product ideas at very early stages of market research. Such encounters can be centered around visual, physical, tactile, kinesthetic, and realistic interactions for the user to respond to. Instead of unreliable mental models from verbal descriptions, utilizing virtual and physical representations results in more useful and valid responses.

TOOLS, TACTICS & TALENT

- > *Sophisticated visualization/simulation media tools*
- > *Highly interactive user focus groups*
- > *Iterative concept simulation process*
- > *Use of specialized visualizers/simulators*
- > *Cross-functional query and evaluation teams*

RESULTS & BENEFITS

- > *Early identification of desired product features*
- > *Better defined user requirements/wants/needs*
- > *Reduced schedule due to early definition*
- > *Minimized late changes or errors*
- > *Increased target market success*
- > *Maximized customer satisfaction*

TWO APPROACHES TO EARLY MARKET RESEARCH

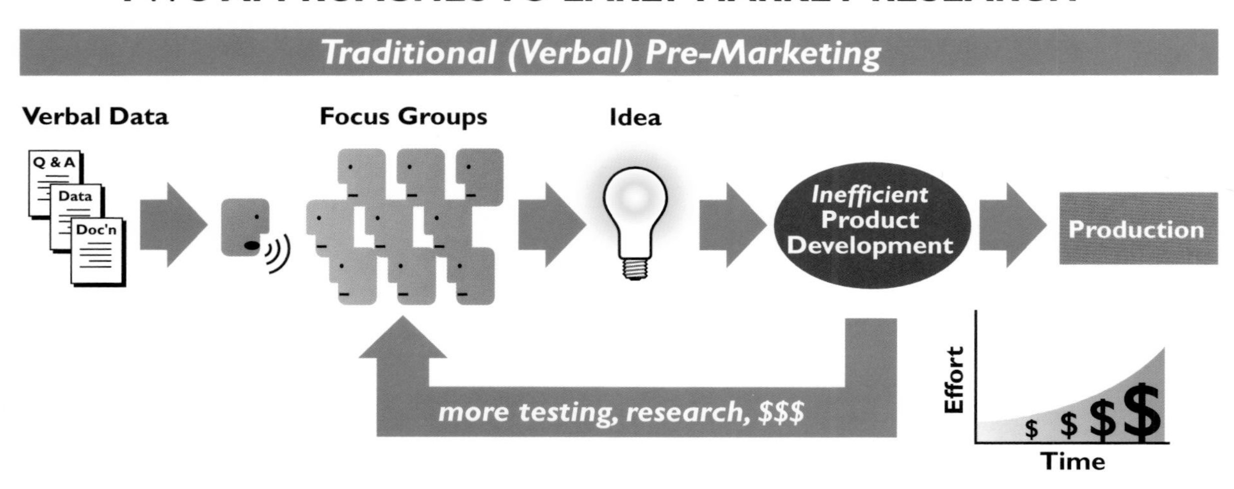

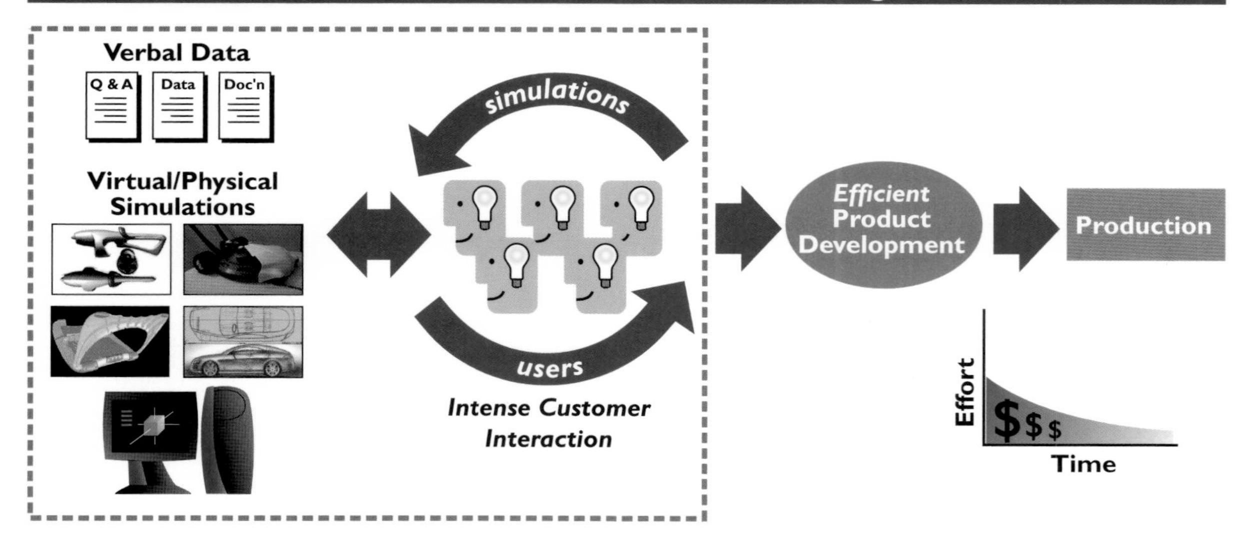

NOTES:

BEGIN
MGMT & FACIL
INNOV & DESIGN
MFG & OPS
BUS & MKTG
END

Digital Pre-Marketing: Using virtual techniques to focus and direct customers early

Many virtual simulation tools are available to the marketing specialist and the product developer to use in creating realistic and interactive imagery for early market studies. Gray Holland, an expert in digital visualization and simulation techniques, uses them to promote product ideas before they have physical embodiment. Using computer-aided industrial design simulation and dynamic animation software, Gray can produce products and their environments so realistic that one cannot tell them from a photograph or video of an actual situation. He uses these simulations and animations to present product variations, scenarios, and environments to his clients so that the product market and its features can be evaluated early in the conceptual phase of development.

Gray Holland: Founder/Principal, Alchemy, San Francisco, California

Gray believes that the transformation of an idea should lie in the hands of its creator. A strong advocate of digital industrial design, Gray is the owner of Alchemy and vice president of digital design innovation at frogdesign in San Francisco. He received a B.S. in transportation design from Art Center College of Design. Though educated at a premier industrial design school, he has a strong background in math and physics, enabling him to successfully integrate designs with engineers. Gray has designed such products as the GM EV-1 electric car, the SGI O2 computer, and several Nike eyewear products.

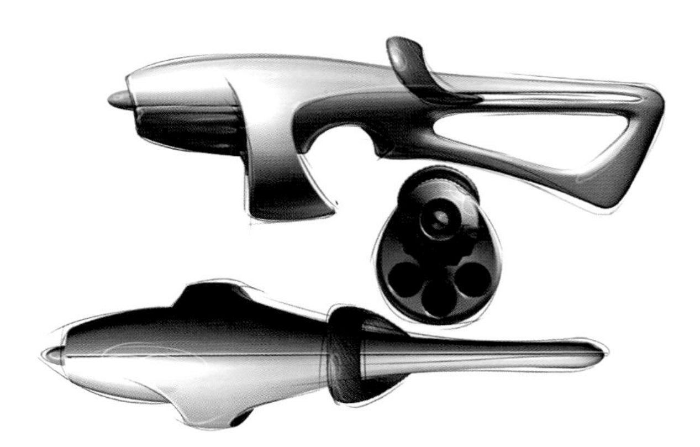

Initial virtual concept ideation for review

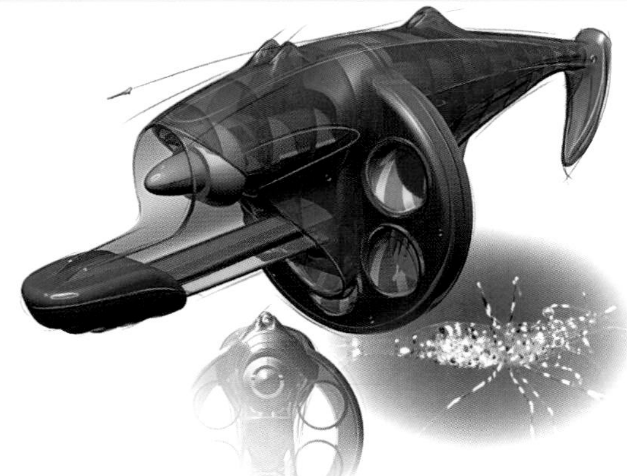

A more refined and detailed concept

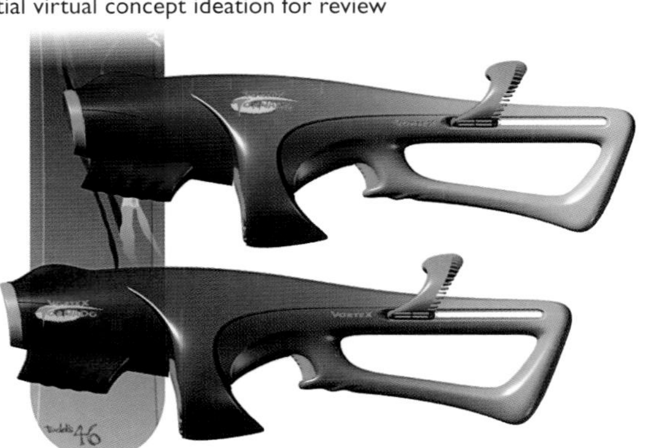

Early product color variation studies

Product packaging design and development

NOTES:

"Great spirits have always encountered violent opposition from mediocre minds."

Albert Einstein

Properly managed product development can become a company's identity

"You are what you eat" applies in the corporate world: your company is what you produce. Give people poorly designed, under-performing products, and you will be perceived as a poorly managed company with an appropriately undervalued stock price. Indeed, it is possible to take this maxim one step further. The design process can actually shape the company. Innovation, after all, tends to permeate the organization in which it occurs, particularly if that company values and celebrates the innovation process. Some companies not only encourage innovation, they manage the design process in order to shape the company and its identity. The process of identifying and meeting the perceived needs of customers not only drives these companies' product development efforts, but those efforts in turn design the company.

TOOLS, TACTICS & TALENT

> *Innovation-oriented corporate management*
> *Investment in innovation people and processes*
> *Cross-functional involvement of ALL departments*
> *Front-loaded, early development efforts*
> *Holistic development: include everyone*
> *Innovation commitment in ALL corporate areas*

RESULTS & BENEFITS

> *Consistent need-meeting innovative products*
> *Smooth development process with fast TTM*
> *Improved product performance in marketplace*
> *Improved brand and company perception*
> *Properly managed, strong corporate identity*
> *Higher equity valuation*

SoMA, Inc.: 514 NW 11th Avenue, Suite 209, Portland, Oregon 97209-3227

SŌMA

SoMA is a strategic product design and development firm with a passion for making objects and technology better serve the people who use them. Their product research, design methodology, and manufacturing logistics help Pacific Rim clients conceive and develop insightful products that promote, revitalize, or transform their organizations. SoMA has developed products for such prominent companies as Nike, Intel, IBM, Carl Zeiss, and Herman Miller.

Corporate Identity

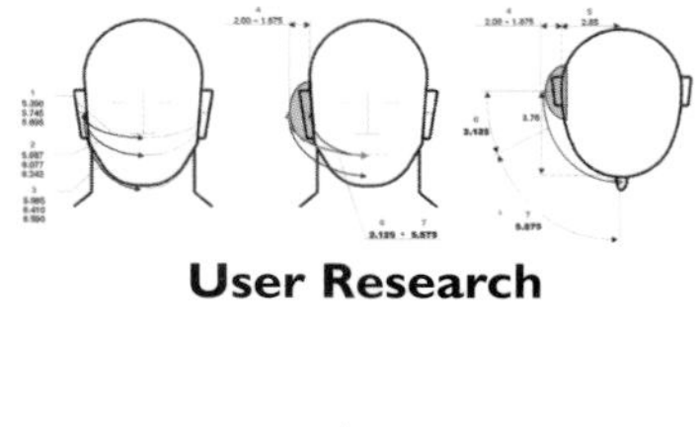

User Research

Design & Engineering

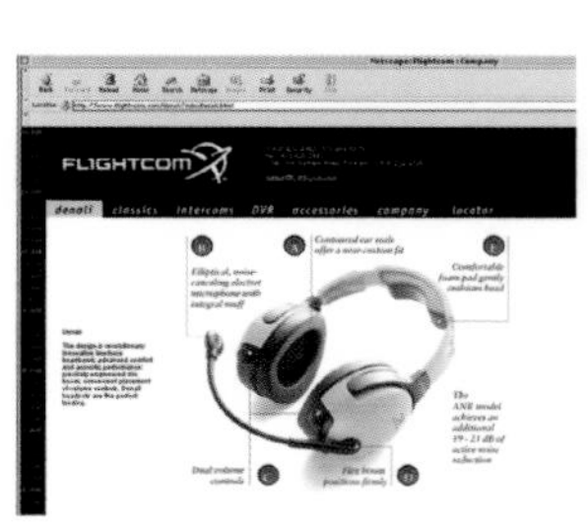

Web site & E-business

Advertising & Promotion

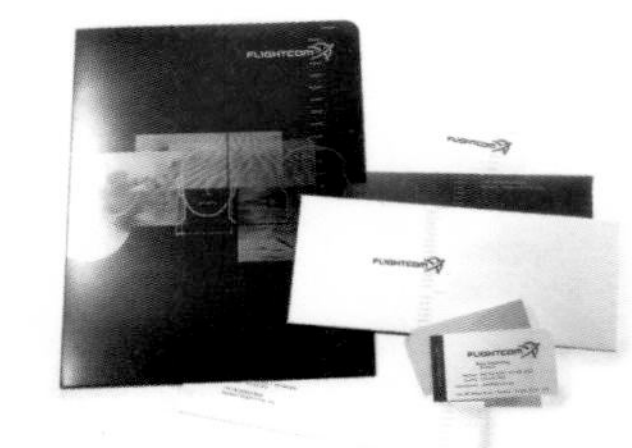

Sales Collateral

Future Products and Opportunities

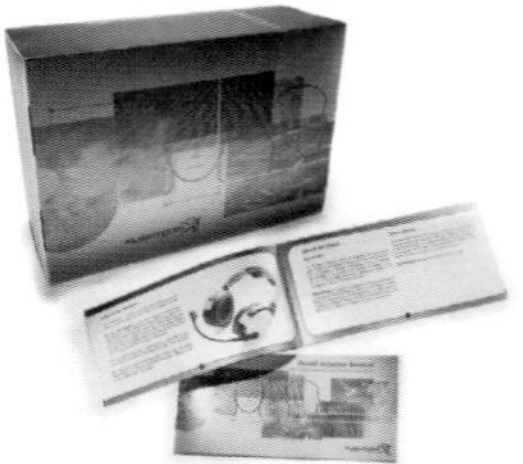

Product Packaging

NOTES:

Product Design Power: Starting from a product need and ending with a new corporate system

If products are truly designed to meet the real needs of users and customers, then those products and their development should drive other aspects of the enterprise, not vice versa. Marketing, packaging, advertising, corporate identity, and sales should evolve from the nature of the product being presented. This happened in the case of SoMA and their work for FlightCom, Inc. The industrial design and product engineering of FlightCom's new headset system drove these other aspects of the company and literally helped transform the entire corporate enterprise around the new design. It is important to note that this was done very early in the process, literally as the product was evolving and being developed. The new product design drove and evolved other areas, resulting in a total enterprise paradigm shift around a new product line. The results: a phenomenally successful re-energized corporate enterprise.

Stevan Wittenbrock: Founder/CEO/President, SoMA, Inc., Portland, Oregon

Steve creates the vision for SoMA and manages the corporate culture. He believes the best design solutions are born when the customer's needs, not technology, drive the development process. He designs to clarify, to order, to question, and to explore his passion. With more than twenty years of design experience, Steve's work has consistently earned recognition from his clients and their customers as well as awards from professional organizations like IDSA and Communication Arts. Prior to pursuing industrial design, Steve studied architecture at the University of Washington. He holds B.S. and M.S. degrees in design from the California Institute of the Arts.

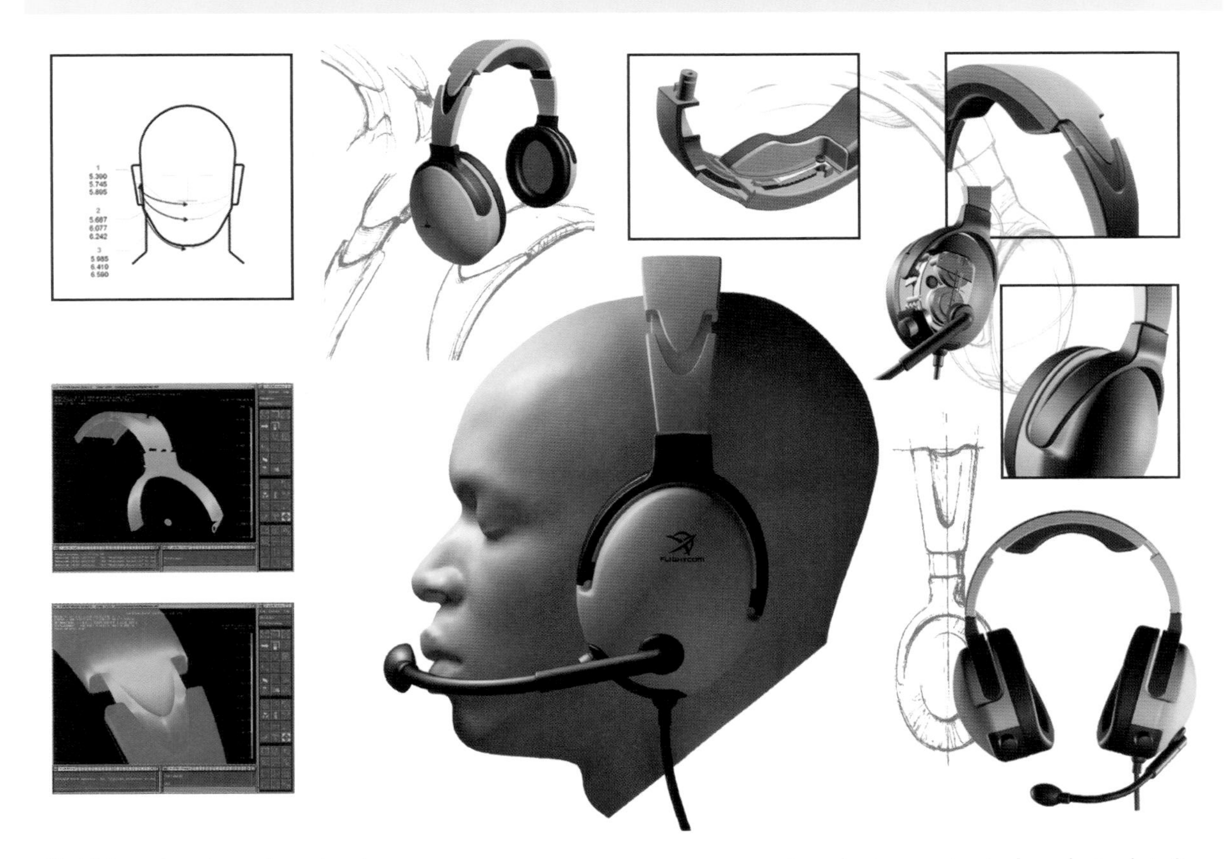

Multi-functional product development: concept, appearance, ergonomics, electromechanical, analysis, detailing

NOTES:

> *"In the beginner's mind there are many possibilities, but in the expert's mind there are few."*
>
> Shunryu Suzuki

USER AS DESIGNER

Engaging the customer in the design process for optimum success

The ideal product would meet the exact needs and wants of the user. Generally, designers must anticipate, estimate, and extrapolate these criteria through their experience, sensitivity, and observation. However, if there were a way to tap into the user's needs directly, products would be better designed. This is often a difficult task due to the limitations of communication, user design abilities, and user availability. Developing tools, tactics, and talent for not only drawing out the desires of the user but to also engage them as designers can be extremely valuable. Techniques such as focus groups and one-on-one interactions that use extensive simulation/interaction methods and tools can accomplish much toward optimized product usability. The Fitch Make-Do-Say process is an excellent example of such a technique.

TOOLS, TACTICS & TALENT

- > *Non-threatening engagement of users*
- > *Provide positive creative environment*
- > *Stimulate ideation and creativity*
- > *Encourage users to explore freely*
- > *There are no failures or mistakes*
- > *Provide extensive simulation exercises*

RESULTS & BENEFITS

- > *Direct engagement of user as product designer*
- > *Optimized design for usability*
- > *Improved product marketability and sales*
- > *Best-in-class product features*
- > *Avoidance of product feature errors*
- > *Competitive market edge in innovation*

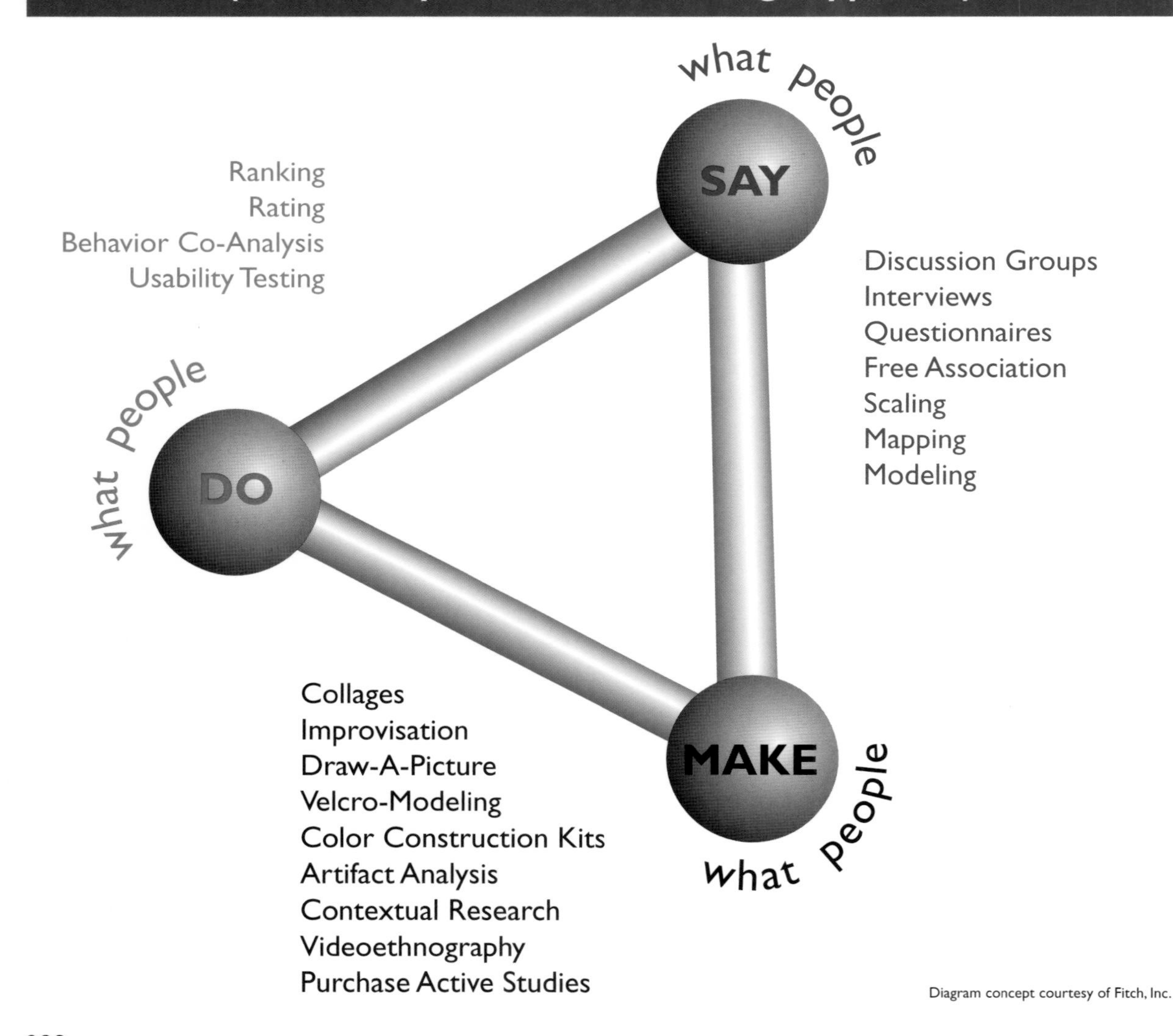

Diagram concept courtesy of Fitch, Inc.

NOTES:

Make-Do-Say: Devices and techniques to assist the user in valuable design input

One way to incorporate the user as designer is to provide interactive sessions using physical items of engagement that will permit them to directly design products. Fitch, an international industrial design and product development consultancy, has done this successfully with their Make-Do-Say process. They have developed a sequence of events that their user groups participate in that includes various stages of interaction. The user develops their own materials relative to their wants and needs. One of the important phases of this process is where a large assortment of soft foam, fabric-covered geometric shapes plus an assortment of small attachments that stick to the main shapes is provided to the user. They can use this object system to create product configurations by selecting certain large forms and attaching the smaller elements that might represent buttons, displays, controls, etc. The overall Make-Do-Say process permits Fitch to engage users in a variety of ways and observe them in what they make, do, and say about the products that they want.

Fitch, Inc.: 10350 Olentangy River Road, P.O. Box 360, Worthington, Ohio 43085

Fitch is one of the largest business and design consultancies in the world and provides expertise in a variety of specialized areas. It focuses on providing its clients with sophisticated product development from conceptualization to marketability. It serves new companies and revitalizes already established ones. Fitch was established by the consolidation of Fitch & Company and RichardsonSmith and became a part of the Lighthouse Global Network in 1999. It is now privately held, has represented such companies as Compaq, IBM, Iomega, Lexmark, and 3M, and is currently participating in a joint venture with GE Plastics.

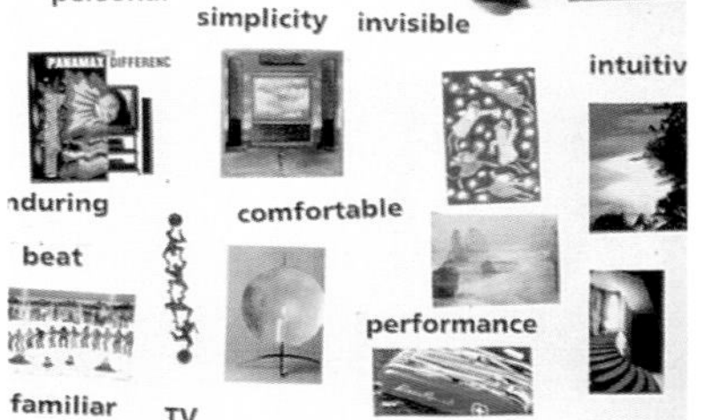

User-developed image board of features

Assortment of idea-configuration elements

Children developing user image boards

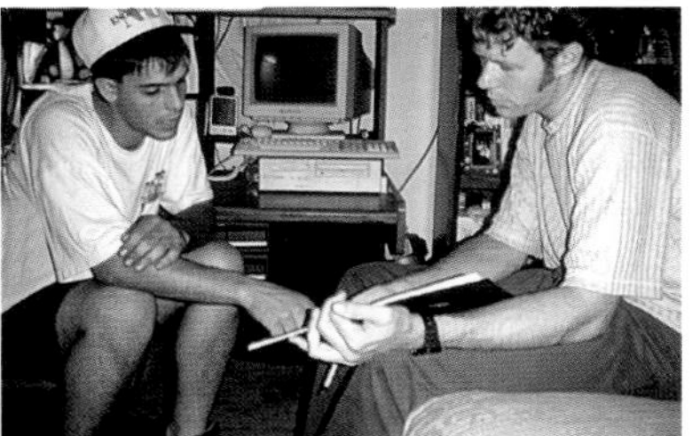

One-on-one interaction with users

Configuration elements plus attachments

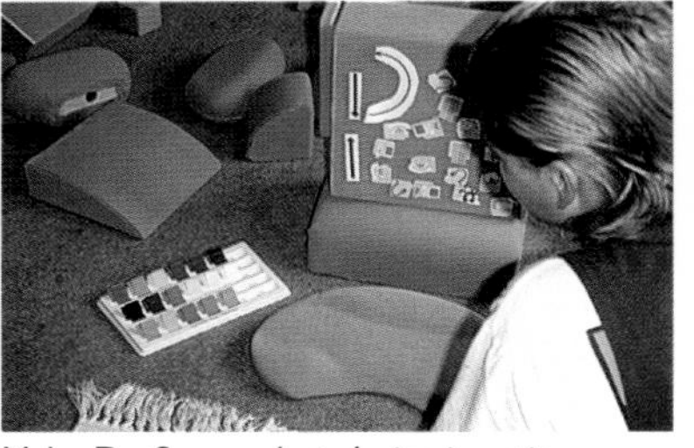

Make-Do-Say product design in action

Interactive tools for user engagement

NOTES:

BUILDING INCUBATORS

"Status is who comes up with the best ideas. It's not who's the oldest or who has the biggest title or who's been at the company the longest."
David Kelley

Developing environments where innovation can be seeded, nurtured, and profitably grown

As many entrepreneurs know, good ideas are often much easier to come by than resources to start development. Even the technologies and processes associated with launching an idea are not always completely known to the idea creator or entrepreneur. The concept of a business incubator can help resolve much of these difficulties by providing the environment, support system, and funding to get a new idea and company successfully off the ground. Incubators can take many forms and can apply to any segment of the development enterprise and community. However, there are certain key features that characterize the most successful ones. With the ultimate objective being to launch new product enterprises that are highly profitable and contribute real value to society, it is imperative that the incubation process is well managed and executed.

TOOLS, TACTICS & TALENT

- *Optimum facilities with latest technologies*
- *State and local government support/funding*
- *Knowledge and resource base ready*
- *Business/management/legal expertise provided*
- *Expertise and resources traded for equity*
- *Leveraged collaboration of tenant entrepreneurs*

RESULTS & BENEFITS

- *Insures new market readiness and viability*
- *Collaborative synergy spawns more innovation*
- *Allows more focus on innovation and creativity*
- *Increases local economic, labor, and tax growth*
- *Creates win/win situation for all participants*
- *Increases investment by outside institutions*
- *Creates benchmark knowledge for development*
- *Develops new business for existing suppliers*

Trudi Lynn Luebberst: Business Manager, SDRC, Milford, Ohio

Trudi Luebberst is involved in several key areas of marketing at SDRC and has developed and managed marketing programs designed to impact sales revenue. She has managed projects in a variety of areas including marketing planning, business development, product release, and vendor development. Trudi continues to work collaboratively with BIO/START, a startup incubator that assists entrepreneurs with new ideas, by her support with SDRC software tools and industry information. Trudi received her B.A. in business from the University of Cincinnati and has over fifteen years of experience in marketing, business, and design collaboration.

NOTES:

Incubator in Action: A biomedical incubator that has started out on the right foot

There can be any number of variations on the incubator start-up model. Any field where innovation is essential is ripe for this tactical tool. These endeavors are common in the fields of biomedical products, business services, and high-technology products and are often located in such areas as Silicon Valley, Los Angeles, Boston, or Austin, Texas. However, any area where product innovation is important and resources are available is viable. Represented on this page is an incubator start-up organization that, while still early in its existence, has many positive attributes. BIO/START is located in the heart of a biomedical development area and university medical school campus and has an excellent facility, a viable business model, competent leadership, surrounding technical and business resources, and a multitude of support structures and resources to help it succeed.

Carol J. Frankenstein: President, BIO/START, Cincinnati, Ohio

Carol Frankenstein is especially interested in the multidisciplinary team approach to product development. Carol received a B.S. from Ohio State University and an M.B.A. from the University of Chicago. She has been director of the Southern Ohio Edison Biotechnology Center and vice president of administration and finance at the James N. Gamble Institute of Medical Research in Cincinnati, Ohio. She is currently president of BIO/START and is working to create an entrepreneurial environment and business support structure for biomedical start-up companies.

Lobby and reception area

Hallway corridor and space connecting labs, services, and offices

Common support services room

Central conference room

Computer and CAE facilities

Typical office space and furnishings

Interconnected wet laboratories

Typical wet laboratory accommodation

NOTES:

TRAINING THE ENTIRE BRAIN

Whole-brained "ambidextrous" thinking is required in today's competitive marketplace

"In the 1960s, researchers discovered that certain traits were resident in each half of the brain when they isolated the hemispheres. It is now widely believed that emotion, intuition, and spatial comprehension reside in the 'right brain', while reason, logic, and analysis are in the left. Curiously, math prowess is a right-brained activity as it involves spatial comparisons. Today's educational methods are heavily geared to the left-brained individual: lectures, reading, and writing are typical. The right-brained individual learns differently, however. Pictures, demonstrations, and experiential exercises are methods that they can embrace and learn from. They frequently draw in their textbooks during lectures, more often are left-handed, and feel differently about themselves than the mainstream. And, as mentioned before, they are often good at math. However, it is not a hard distinction between the two groups. These qualities exist in different measures in all of us. Our goal, then, as responsible trainers and educators, should be to train whole-brained, balanced individuals."—Marty Smith

TOOLS, TACTICS & TALENT

- > *Use both right- and left-brained styles of teaching*
- > *Strive for whole-brained ambidextrous thinking*
- > *Exercise and develop less-prominent side of brain*
- > *Develop left/right brain communication skills*
- > *Train to be able to switch brain sides at will*

RESULTS & BENEFITS

- > *Better overall thinking and performance*
- > *Improved creativity and innovation*
- > *Better communication skills to all audiences*
- > *Cross-functional understanding and cooperation*
- > *Improved learning capabilities and knowledge*

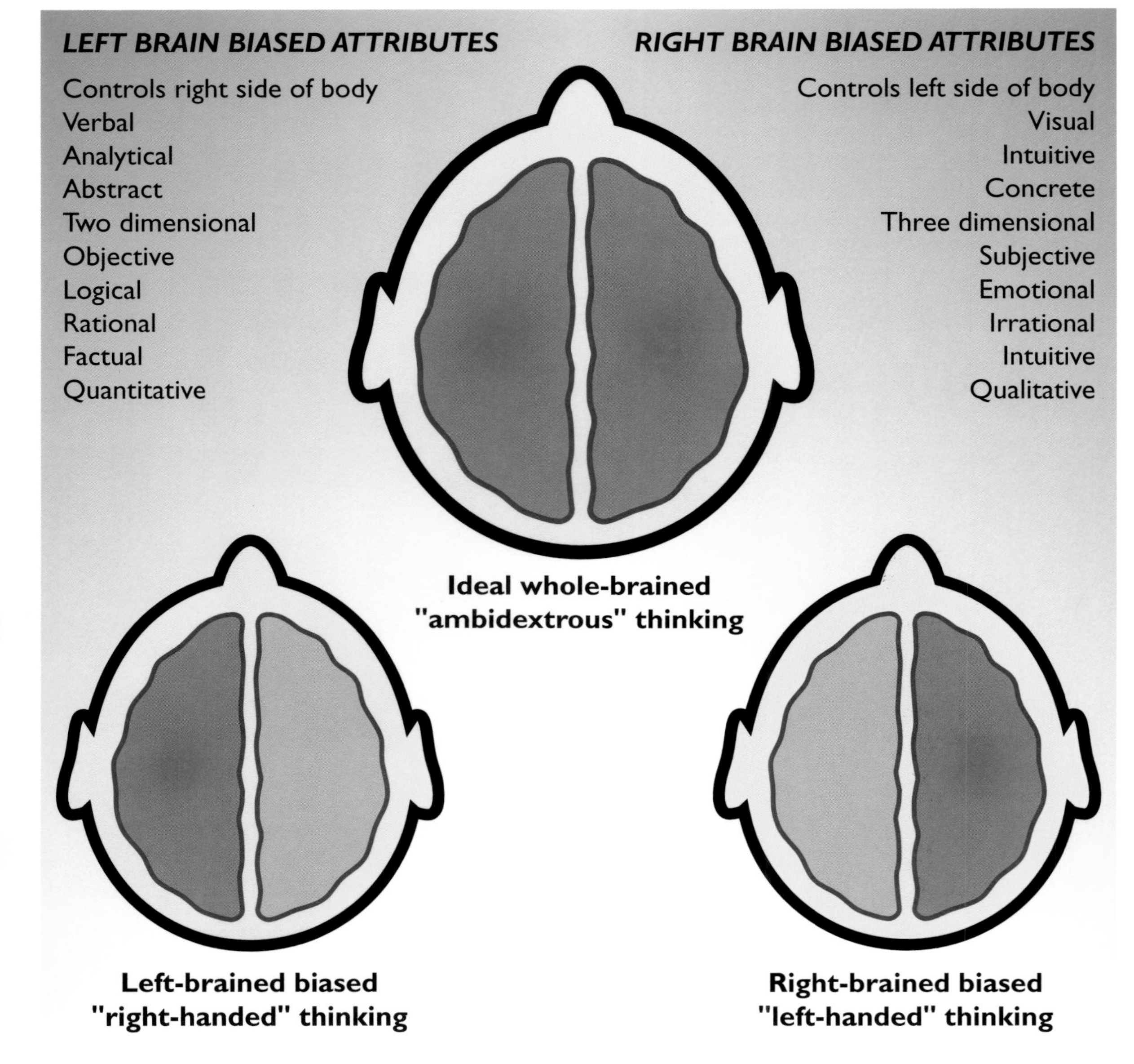

NOTES:

Right-Brained Training: Developing the "other" brain appropriately and to full potential in everyone

"The human being is an amazingly adaptable creature! How else to explain why we painfully endure dripping coffee pots, blinking VCRs, and other annoying devices. But even more amazing is how right-brained students have adapted to a primarily left-brained world. For over fifteen years I have observed the qualities that identify certain kinds of people that will be more creative than others and how they use those qualities to adapt. I am reminded of the characters in the film, *The Breakfast Club*. The art student was this spacey loner who turns out to be more like the jock, prom queen, or hoodlum than imagined. The outsiders who set themselves apart because they know they are different but don't really know why. This is due to the different ways in which we learn and whether we favor the left or right side of our brain. We all have creative potential. What is needed are learning environments that nurture that creativity and give the student the tools to communicate it. Developing the right side of the brain is essential to maximum creativity and innovation in all of us."—Marty Smith

TOOLS, TACTICS & TALENT

- > *Emphasize 2D and 3D visualization*
- > *Develop drawing and sketching capabilities*
- > *Develop 3D modeling capabilities*
- > *Drive intuitive, kinesthetic, and visual thinking*
- > *Develop directed fantasy skills*
- > *Provide highly stimulating sensory environments*
- > *Training to be interactive and stimulating*

RESULTS & BENEFITS

- > *Developed and improved right-brained thinking*

C. Martin Smith: Chair, Product Design, Art Center College of Design, Pasadena, California

Marty has been chairman of the product design department at Art Center College of Design in Pasadena, California since 1991. He received his B.F.A. from the Cleveland Institute of Art in 1972 and after co-owning and managing a successful product design consultancy, he became Director of Automotive Projects for Designworks/USA. There he created a substantial presence in the automotive industry with notable work for Chrysler, General Motors, Volvo, and Subaru. Marty continues to consult with major companies such as Coca-Cola, Reebok, Olympus, and others. He also speaks internationally on a variety of contemporary design issues.

Design studio/laboratory environments

Two- and three-dimensional concept visualization

Collaborative design projects

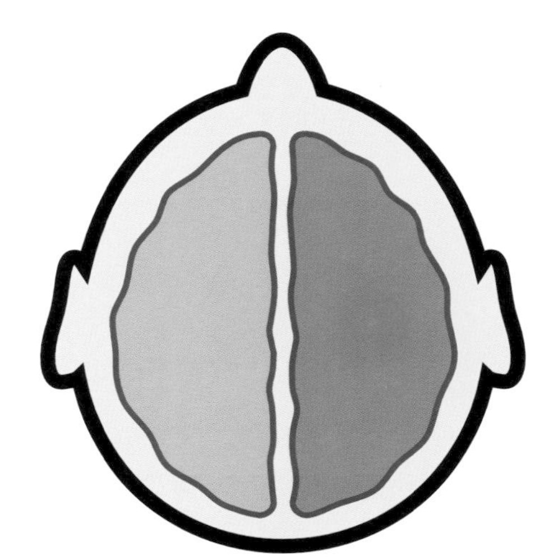

Overall environment for creativity

Interactive trainers

Environments, tools, tactics, and talent for training the right brain

NOTES:

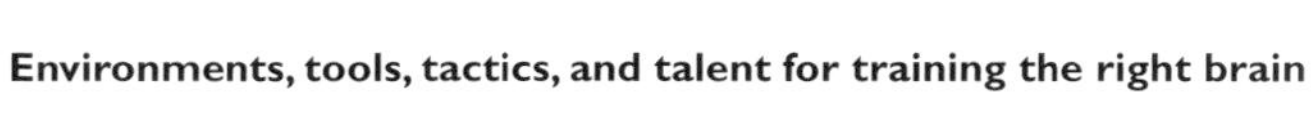

Training Via Consultants: Innovation and creative process education from outside resources

A quite viable avenue for corporate training and education is through product development consultancies that provide seminars and educational media such as IDEO. Its education component is called IDEO University and is headed by Barry Katz. IDEO U provides seminars and training sessions focused on the creative process and innovation in product development. A company may retain them to put on a half-day to a several-day set of sessions that will enlighten and stimulate designers, engineers, managers, and other corporate personnel. IDEO U's presenters facilitate highly interactive sessions using hands-on projects to engage participants directly. These are educational as well as fun team-building exercises. The projects range from simple brainstorming and idea-generating exercises to the actual design of a product using sketching, models, and computers.

TOOLS, TACTICS & TALENT

> *Identification of consultant training*
> *Testing and qualification of training resources*
> *Variety and flexibility of outside training*
> *Flexible options for in-house vs. outside*
> *Ease of attendance of staff to any training*
> *Management participation in training*

RESULTS & BENEFITS

> *New and innovative creative processes*
> *Specialized training information and knowledge*
> *Unbiased expertise and encouragement*
> *Staff stimulation and rejuvenation*
> *Improved product development results*
> *Better team interaction and performance*

Barry Katz: Dean, IDEO University, IDEO, Palo Alto, California

Barry Katz is both an IDEO Fellow and Dean of IDEO U, which conducts several Innovation Workshops for industry. When Barry is not serving in an IDEO capacity, he is a professor of design at the California College of Arts and Crafts in San Francisco and a consulting professor in the School of Engineering at Stanford University. Barry was educated at McGill University in Montreal and the London School of Economics and received his Ph.D. from the University of California. He has published dozens of articles in a variety of journals and is a contributing editor to several magazines.

Sharing a new idea

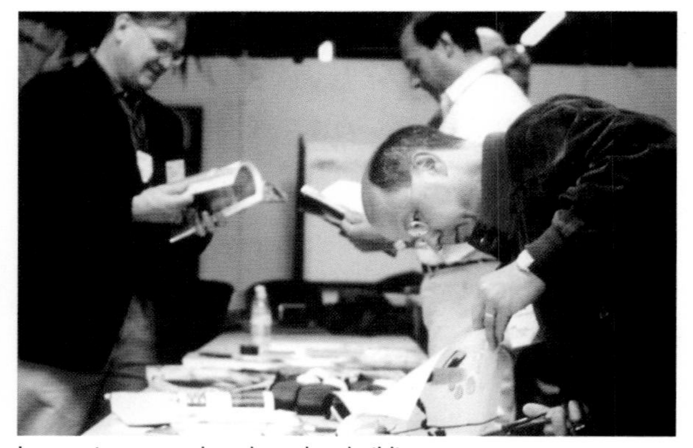

Interactive research and mockup building

Collaborative concept development

Enjoyable and stimulating brainstorm session

NOTES:

Industrial Strength Training: Creating a corporate "college" to train for exact capabilities required

A common challenge for high-technology development organizations is finding, hiring, and retaining highly qualified technical personnel. Often a basic four-year technical college education does not provide the necessary level of skill that a company needs from its designers and engineers. Frequently both recent graduates and experienced technologists need additional training in development functions with the latest tools and techniques. Lexmark has creatively addressed this issue by developing a custom-designed in-house program. They find high-potential recruits and train them in required CAE skills, tools, and processes over a one-year period that includes pay. Those who achieve Lexmark's standards during this one-year training period are poised for an offer of a permanent position at Lexmark. Via this successful program, Lexmark is able to staff critical development areas with competent proven performers specifically trained for success.

TOOLS, TACTICS & TALENT

> *Recognition of specialized training needs*
> *Commitment to high-quality personnel*
> *Dedicated training manager/facilitator/mentor*
> *Clearly defined training strategy and tactics*
> *Adequate facilities, tools, and processes*
> *Appropriate evaluation process*

RESULTS & BENEFITS

> *Optimized development talent pool*
> *Maximized design expertise and skills*
> *Insurance of competently trained staff*
> *Best application of tools and processes*
> *Recruitment and retention of best/brightest talent*
> *Staff competition and motivation to excel*

Lexmark International, Inc.: 740 West New Circle Road, Lexington, Kentucky 40550

LEXMARK™

Lexmark is a global developer, manufacturer, and supplier of printing solutions products including laser, inkjet, and dot matrix printers and associated supplies for the office and home markets. It was formed in 1990 by a private investment firm in connection with the acquisition of IBM Information Products Corporation from IBM. It has won numerous business and design awards for its products, and to maintain its competitive position it strives to attract, retain, and reward talented employees.

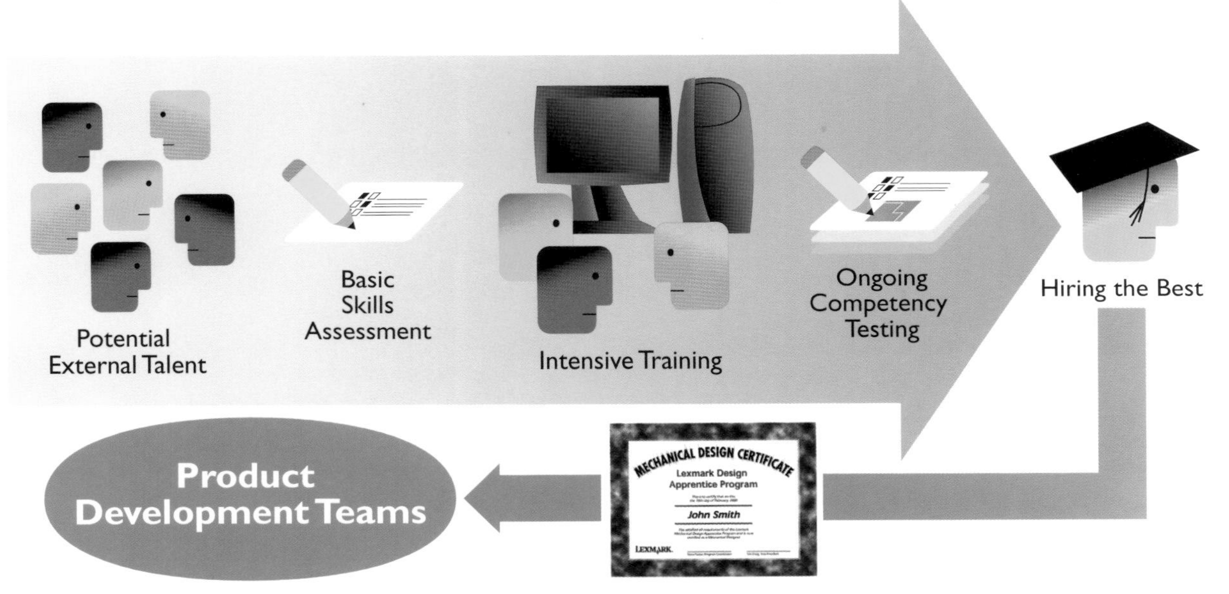

Regular classroom training sessions

One-on-one interaction and support

Focused learning of essentials and techniques

NOTES:

"Ideas must work through the brains and the arms of good and brave men, or they are no better than dreams."
Ralph Waldo Emerson

INNOVATION AND LAW

Addressing the many legal issues surrounding product innovation and intellectual property

Often neglected or left until late in the development process are the many legal issues, relationships, agreements, and associated counsel surrounding a new venture and its intellectual property. Neglecting contracts and agreements, especially early in the process, can result in unfortunate or even disastrous situations later. Whether it is the legal relationship between new partners, a contract for custom development work, specifications of deliverables from a turnkey supplier, or other legal considerations, the proper execution of documents, agreements, and ground rules is essential for success. Retaining competent, experienced, and compatible legal counsel is mandatory. The schematic process on this page indicates a rough plan for approaching intellectual property protection in product development.

TOOLS, TACTICS & TALENT

> *Timeliness in ALL activities and actions*
> *Constant and consistent documentation*
> *Intellectual property identification/clarification*
> *Intellectual property refinement and protection*
> *Managed communication (use caution/control)*
> *Compatible/creative legal counsel and support*
> *Agreements executed between all parties*

RESULTS & BENEFITS

> *Early identification of potential conflicts*
> *Avoidance of later litigation*
> *Security and confidence of relationships*
> *All deliverables/specifications clearly spelled out*
> *Freedom to focus on critical development issues*
> *Peace of mind for all concerned*

John Eustermann: Attorney at Law, Hutchison, Hammond and Walsh, PC, West Linn, Oregon

John is a licensed attorney in Washington, Oregon, and California, and comes from a creative family background involving design, medicine, and film. He practices in the areas of commercial transactions and corporate law. John is interested in merging e-business with product development and has a strong interest in the design disciplines and product innovation. John has a B.A. in philosophy and an M.B.A. in finance from the University of Notre Dame and his Doctor of Jurisprudence from the University of Puget Sound School of Law. He is a member of the Oregon, Washington, and California Bar Associations and the Institute of Certified Managerial Accountants.

1 CREATING
The IDEA!! (Ideas alone are cheap; they must ultimately have real value.)

2 DOCUMENTING
Make idea tangible: write, tape, photo, video, sketch, etc. (Start creating idea value.)

3 RECORDING
Quick protection: send copy of docn. in registered letter to self for record (file, do not open).

4 VALIDATING
Discussion of idea: NDA's presented and signed prior (start value protection).

5 REFINING
Idea value research: resources, web, lawyers, books, papers, etc. (validation of value).

6 PROTECTING
Expert legal advice: attorneys for patent, TM/SM, copyright, contract, license, relationship.

7 DEVELOPING
Design, build, source, process, partner, license, assign, contract, produce, market.

8 PRODUCING
Facilities, delivery, contracts, company, employees, services, competition, cash flow, finance, etc.

9 PROFITING
ROI, revenue, royalties, stock, equity, company sale, product sales, taxes, risks, etc.

NOTES:

Being First Can Be A Problem: Creating design innovations and trends has its legal challenges

Who would think that a mobile phone would share the same piracy problems as a pair of designer jeans! Nokia has realized that good design and recognized trends provide fertile ground for pirates and copycats. Since the introduction of the world's first user-changeable cell phone cover on the 5100 series, Nokia has seen how many unauthorized copies can flood the marketplace. Keeping ahead of these pirates is a challenge that Nokia is faced with every day. There are some unexpected benefits—after-market covers are available in abundance, which promotes personalization as a reason to purchase a Nokia phone. Still, only Nokia OK (licensed after-market covers) and Nokia Original Accessories are made to meet Nokia's stringent quality standards. Pirates often use cheap materials and processes that may adversely affect the durability and performance of the phone. Consequently, Nokia must be ever-diligent in protecting its valuable intellectual property and proprietary designs.

Alastair Curtis: Group Design Director, Nokia Mobile Phones, Nokia Design Center, Calabasas, California

Alastair heads a team of designers in Los Angeles, California responsible for conceptualizing the design and helping to bring to market Nokia's latest range of mobile terminals. Prior to joining Nokia, Alastair spearheaded design work with a number of leading organizations in the U.K. He began his training in the field of design at Brunel University where he received a B.Sc. (with honors) in industrial design, and then at the Royal College of Art where he was awarded an M.A. in industrial design engineering.

The Nokia Design team spends great effort on quality graphics (line, color, image design, and quality). Copies can be identified via inferior quality standards (note the poor graphics found on copies shown below). Toys, which are Nokia look-alikes, are not authorized but, again, show that Nokia designs are popular trend setting solutions.

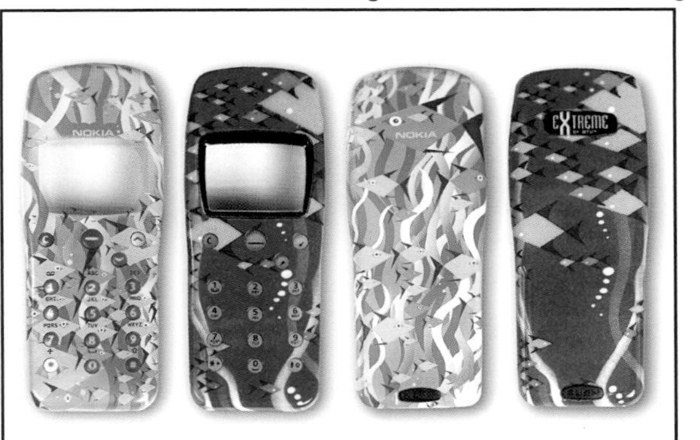

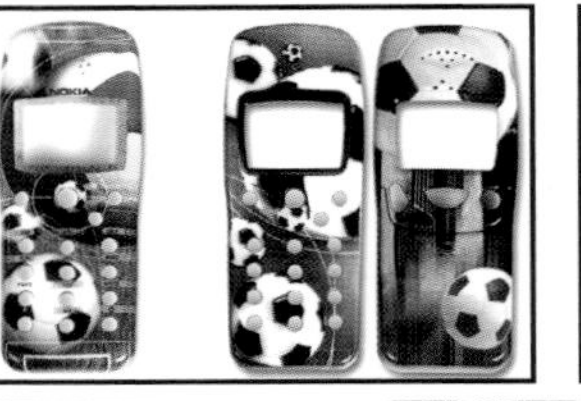

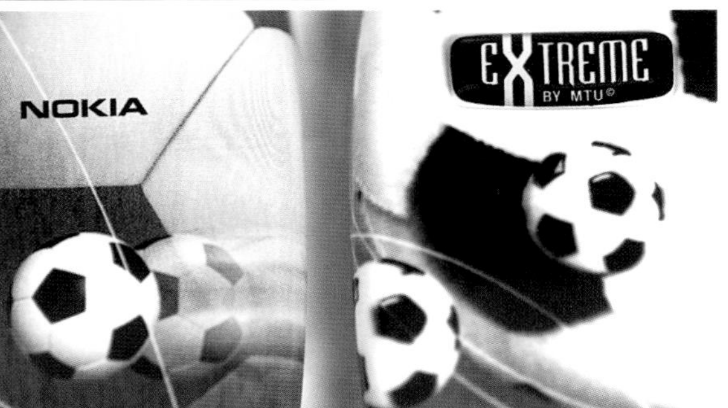

NOTES:

"Sometimes we stare so long at a door that is closing that we see too late the one that is open."

Alexander Graham Bell

"An idea can turn to dust or magic, depending on the talent that rubs against it."

Bill Bernbach

ENDING STUFF

BEGIN
MGMT & FACIL
INNOV & DESIGN
MFG & OPS
BUS & MKTG
END

Some incredible virtual imagery

Gray Holland, Alchemy

Freightliner Corp.

Designworks/USA (BMW)

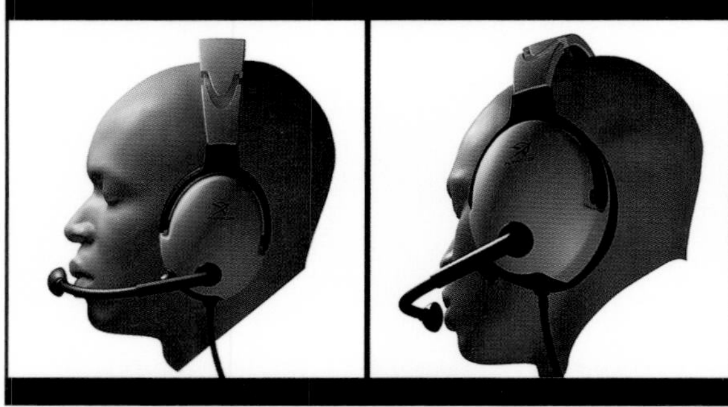

SoMA, Inc.

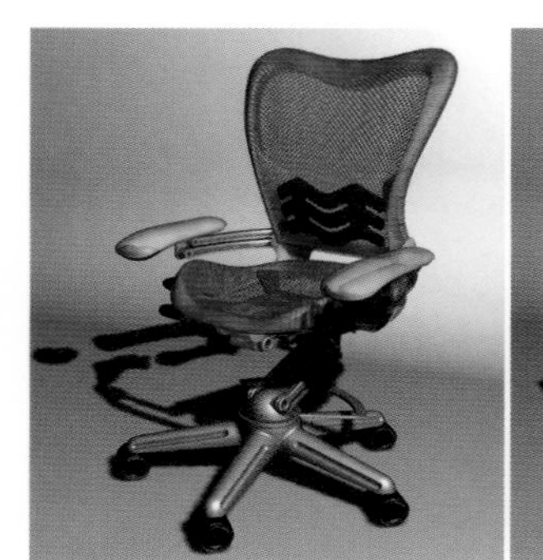

FUSE, Inc.

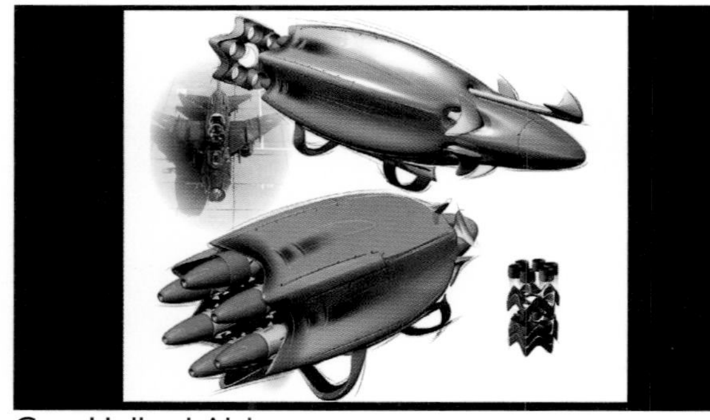

Gray Holland, Alchemy

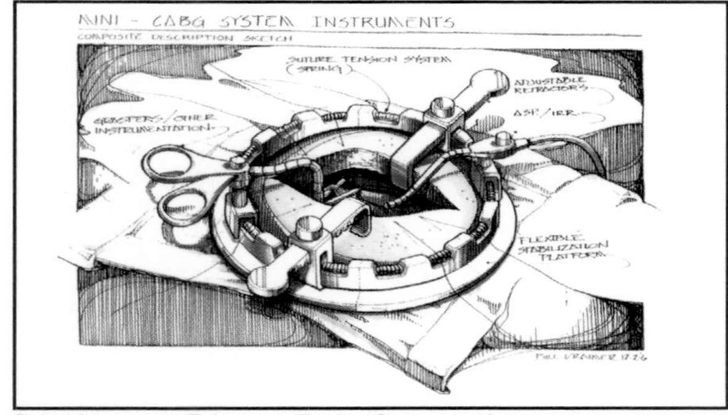

Bill Kraimer, Ethicon Endo-Surgery, Inc.

Curt Croley, Ethicon Endo-Surgery, Inc.

Bill Kraimer, Ethicon Endo-Surgery, Inc.

NOTES:

Gallery (continued)

IAS Design

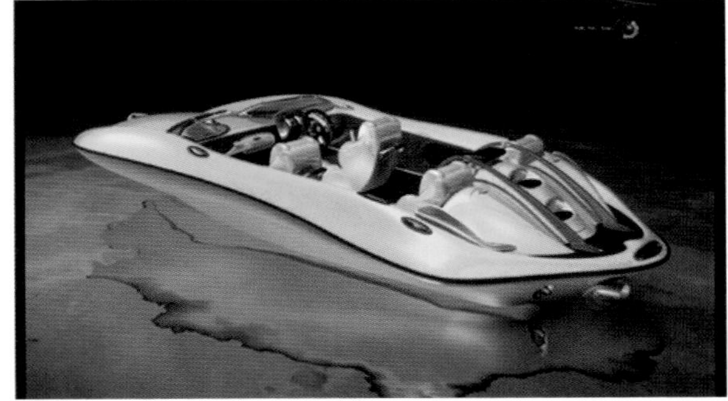

IAS Design

BMW Corporation

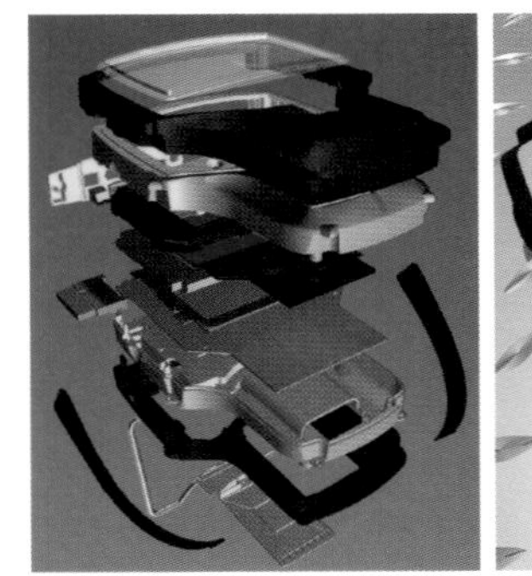

FUSE, Inc.

Ford Motor Company

Walter Dorwin Teague, Inc.

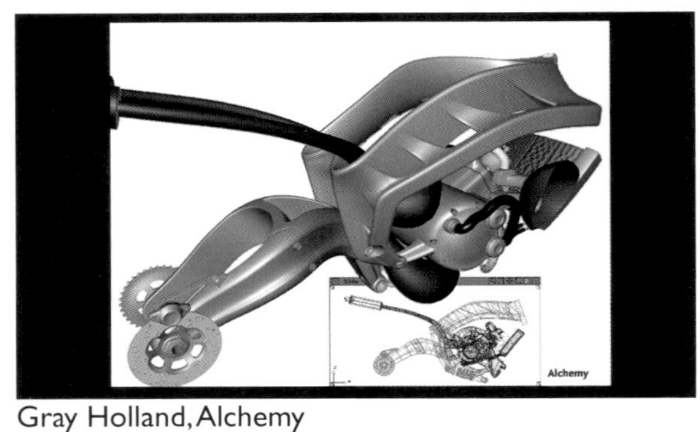

Gray Holland, Alchemy

Habib Zargarpour

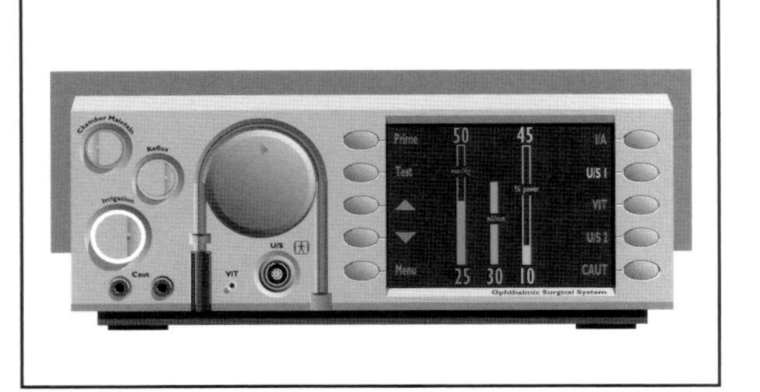

Steve Montgomery, bioDesign, Inc.

NOTES:

Gallery (continued)

Glenn Walters, Brooks Stevens Design, Inc.

Raphael Zammit, Art Center College of Design

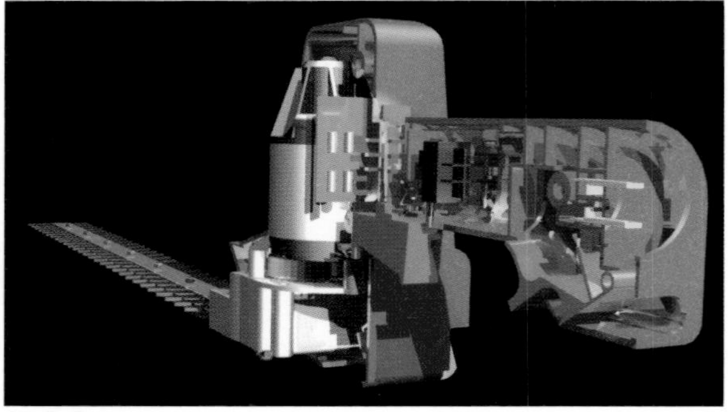
SDRC

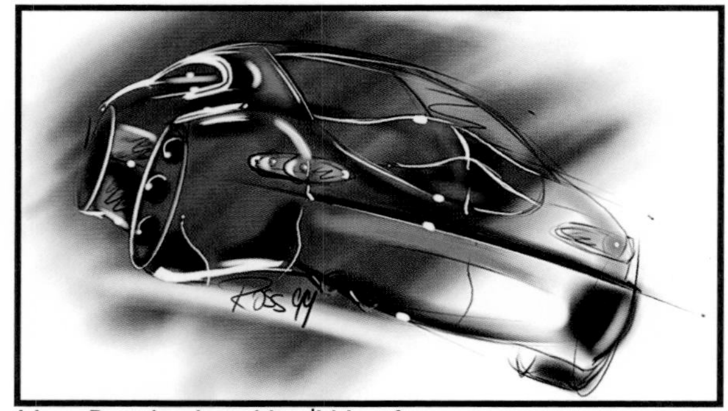
Uwe Rossbacher, Alias|Wavefront

XingXing Magic-UBC

Condit Exhibits, Inc.

Frederick Reber, Eastman Kodak Co.

Ginko Designs, Inc.

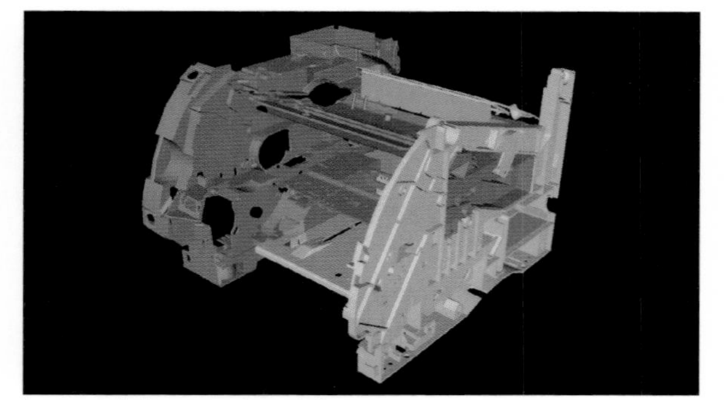
Lexmark International, Inc.

NOTES:

Gallery (continued)

Ian Sands, Art Center College of Design

R. Paul, Alias|Wavefront

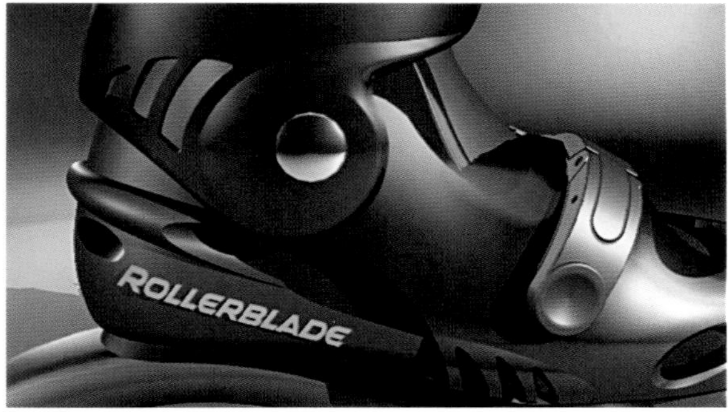

Todd J. Olson, Rollerblade, Inc.

Umit Altun, Peugeot

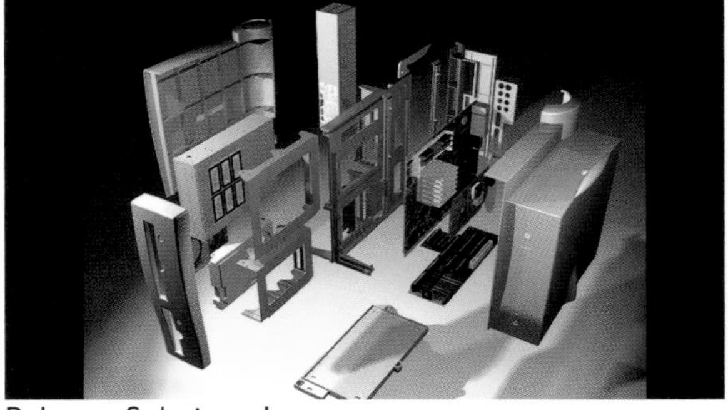
Polymer Solutions, Inc.

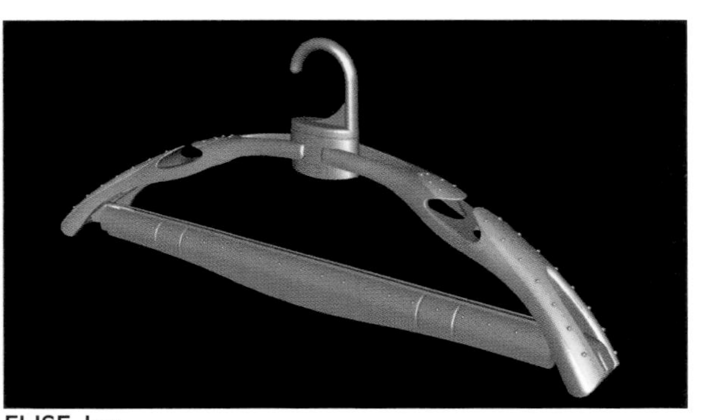
FUSE, Inc.

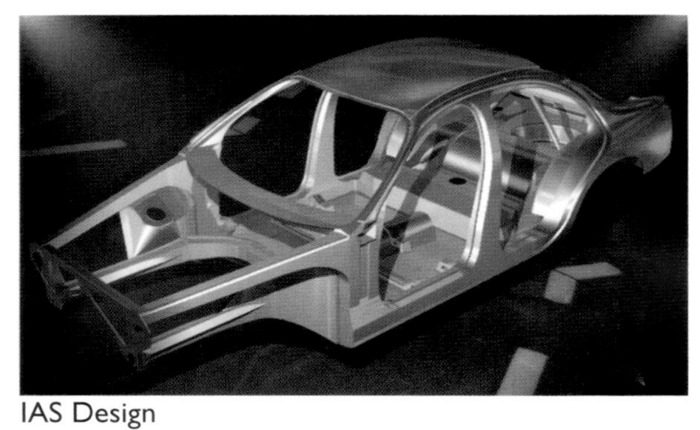
IAS Design

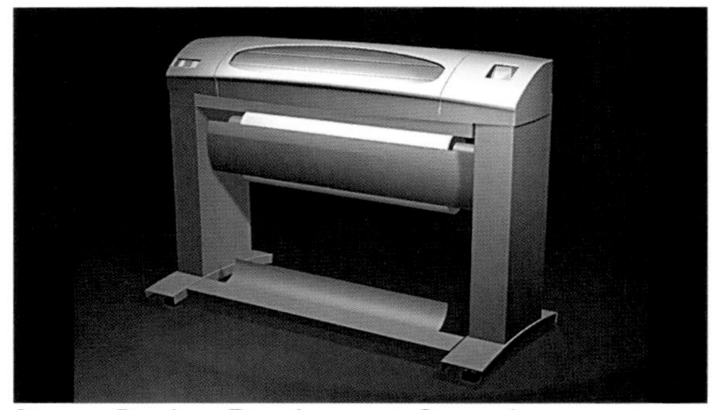
Stratos Product Development Group, Inc.

Maytag Company

NOTES:

Recommended publications addressing innovation and creativity in product development

James L. Adams.
Conceptual Blockbusting: A Guide to Better Ideas.
Third Edition. Cambridge, Massachusetts: Perseus Publishing, 1990.
ISBN: 0201550865

Guy Kawasaki and Michelle Moreno.
Rules for Revolutionaries: The Capitalist Manifesto for Creating and Marketing New Products and Services.
New York: Harper Business, 1999. ISBN: 088730995X

Kurt Hanks and Larry Belliston.
Draw! A Visual Approach to Thinking, Learning and Communicating.
Los Altos: Crisp Publications, 1992.
ISBN: 156052054X

Preston G. Smith and Donald G. Reinertsen.
Developing Products in Half the Time: New Rules, New Tools.
Second Edition. New York: John Wiley & Sons, 1997.
ISBN: 0471292524

B. Joseph Pine II and James H. Gilmore.
The Experience Economy: Work is Theatre & Every Business is a Stage.
Boston: Harvard Business School Press, 1999.
ISBN: 0875848192

Tom Peters and Dean LeBaron.
The Circle of Innovation: You Can't Shrink Your Way to Greatness.
New York: Vintage Books, 1999.
ISBN: 0679757651

Jerry Hirshberg.
The Creative Priority: Putting Innovation to Work in Your Business.
New York: Harper Business, 1999.
ISBN: 0887309607

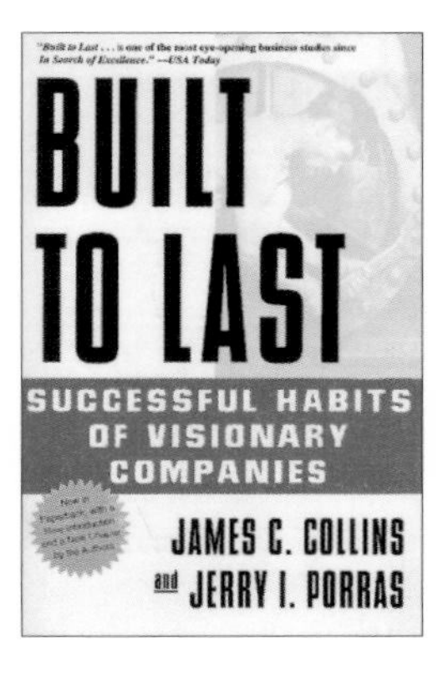

James C. Collins and Jerry I. Porras.
Built to Last: Successful Habits of Visionary Companies.
New York: Harper Business, 1997.
ISBN: 0887307396

Donald A. Norman.
The Design of Everyday Things.
New York: Doubleday Books, 1990.
ISBN: 0385267746

NOTES:

Resources (continued)

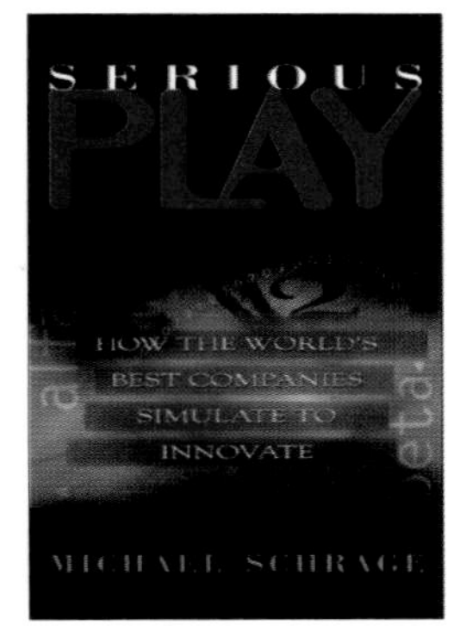

Michael Schrage.
Serious Play: How the World's Best Companies Simulate to Innovate.
Third Edition. Reading, Massachusetts: Harvard Business School Press, 1999.
ISBN: 0-87584-8141

Donald Reinertsen.
Managing the Design Factory: A Product Developer's Toolkit.
New York: Free Press, 1997.
ISBN: 068483991 l

Ray Kristof and Amy Satran.
Interactivity By Design: Communicating with New Media.
Mountain View: Adobe Press, 1995.
ISBN: 1568302215

Tom Peters.
Reinventing Work: The Professional Service Firm 50.
New York: Alfred A. Knopf, Inc., 1999.
ISBN: 0375407715

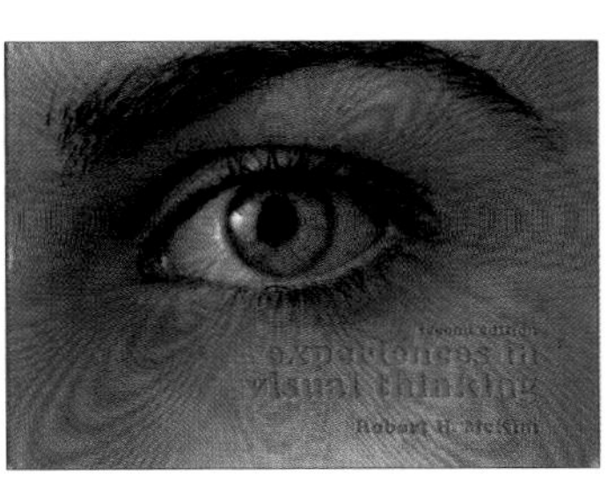

Robert H. McKim.
Experiences in Visual Thinking.
Second Edition.
Boston: PWS Publishers, 1980.
ISBN: 0818504110

I.D.: The Magazine of International Design.
New York: Gigi Grillot.

Gordon MacKenzie.
Orbiting the Giant Hairball: A Corporate Fool's Guide to Surviving with Grace.
New York: Viking Press, 1998.
ISBN: 0670879835

Tom Peters.
The Tom Peters Seminar: Crazy Times Call for Crazy Organizations.
New York: Random House, 1994 (Cassette).
New York: Vintage Books, 1994 (Book).
ISBN: 0679754938

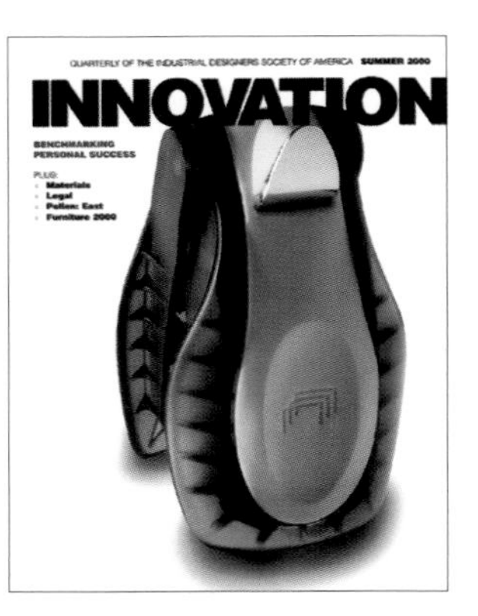

Innovation: The Quarterly of the Industrial Designers Society of America.
Great Falls: IDSA.

NOTES:

ART CENTER COLLEGE OF DESIGN

One of the world's premier industrial design and visual arts schools

Art Center College of Design (ACCD): Industrial Design—Product Design, Transportation Design, Entertainment Design—Pasadena, California

Mark Andersen: Co-Chair, Product Design

Mark is currently assistant department chair of product design at Art Center and also served as acting department chair for one term. Mark is also principal of Direct Design, a consultancy that specializes in entrepreneurial start-up product development programs. Mark received his B.S. in industrial design from Art Center in 1987 and has been a faculty member since 1994. Prior to starting his own company, Mark worked as a senior designer and project manager and has been awarded several design awards including the IDEA Silver and Bronze Awards. Mark has spoken at the IDSA World Conference and has received the Art Center Enrichment Grant for research in design technology.

Geoff Wardle: Director, Corporate Relations

Geoff is corporate relations director at Art Center and is responsible for encouraging businesses and industries to become active contributors to the Art Center educational process by providing their innovative expertise directly to classrooms. Geoff is an experienced automobile designer who has an understanding of creative design, administrative and managerial responsibilities, and educational issues. He has held such positions as chief designer, design consultant, design director, product designer, and transportation design department chairman. Geoff has consulted for and been employed at several international corporations including Abtech 84 Ltd., Chrysler Europe, Ford Asia Pacific, and SAAB Personbildivisionen.

Rob Bruce: Design Professional (alumnus)

As design director at Astro Studios, Rob has worked in the product development community for over seven years with Astro, frogdesign, and IDEO. Most recently Rob led the Astro creative team for Nike's Triax™ line of high performance sports watches. Past design experience includes work for Apple, Coca-Cola, Compaq, Nike, Sony, Samsung, Specialized, and Steelcase. Rob has received design awards from both ID Magazine *and IDSA/*BusinessWeek *in recognition for the Nike Triax™ watches, most notably the Design of the Decade. His work has been published in both consumer and trade magazines nationally and internationally. Rob holds a B.F.A. from Art Center.*

NOTES: ______

Product design and engineering education at its most creative and innovative

Stanford University, Department of Mechanical Engineering, Design Division: Stanford, California

Rolf A. Faste: Associate Professor

Rolf was professor of design in the College of Visual and Performing Arts at Syracuse University before he came to Stanford University in 1984. Rolf received his B.A. in mechanical engineering from the Stevens Institute of Technology, an M.S. in engineering design from Tufts University, and a Bachelor of Architecture degree from Syracuse University. He is especially interested in technical and aesthetic creativity, problem identification, and humanistic considerations when designing products. In addition to being an associate professor and director of the Product Design Program at Stanford, Rolf has written several articles including "The Role of Aesthetics in Engineering."

David Beach: Professor-Teaching

David not only teaches in the areas of product design and manufacturing at Stanford, but he also directs the Product Realization Laboratory there, which allows university students to design, build, and manufacture their product. David received both his B.S. and M.S. in engineering (product design) from Stanford University. He teaches machining techniques and methods at Stanford and is also program director of Manufacturing Systems Engineering (MSE), which offers a professional degree aimed at initiating individuals into practical manufacturing processes.

Kosuke Ishii: Associate Professor

Kos received his B.S. in mechanical engineering from Sophia University in Tokyo, his M.S. in mechanical engineering from Stanford University, his Master of Engineering in control engineering from the Tokyo Institute of Technology, and his Ph.D. in mechanical engineering from Stanford University. Kos is currently the chair of the ASME Computer and Information in Engineering Division and is an associate editor of the Journal of Mechanical Design (ASME) and AI in Engineering. He has written numerous articles on "robust design" and "life-cycle engineering" and has received many awards, including the GM Outstanding Long Distance Faculty Award (in 1996).

The Stanford Product Design Program

This program, a unique blend of engineering and art education affiliated with Stanford's art department, was established in 1958 by Prof. Robert McKim in the Design Division. Emphasis is on basic mechanical engineering and aesthetic design while providing the knowledge to develop projects from concept to functional prototype. The program's course work focuses on the need for creating products that will improve society by encouraging students to utilize their creativity, imagination, and technical knowledge. Applicants are required to possess both an art and design portfolio and a unique technical awareness. The program offers the B.S. and M.S. degrees and graduates often find employment in well-known high-technology corporations and firms or start their own entrepreneurial businesses.

The Stanford RPM Programs

The Design Division has two innovative facilities for quickly transforming student concepts into reality via product realization and rapid prototyping. The Product Realization Lab (PRL), started by Prof. David Beach, offers prototyping capabilities to students and has spaces dedicated to foundry, machining, plastics molding, welding, and woodworking, while providing the latest computer aided drawing, manufacturing, and prototyping systems. In addition, the PRL has a Student Design Loft, where students are encouraged to develop and assess their design developments. The Rapid Prototyping Lab (RPL) is devoted to improving the efficiency of rapid prototyping technology and the quality of design by reducing design cycle times and encouraging student designers to experiment before making final decisions.

Manufacturing Modeling Laboratory

MML is a Stanford Alliance for Innovative Manufacturing (AIM) initiated laboratory located in the new Thornton Engineering Center. The lab serves as a repository of manufacturing models as well as a focus of research on design and manufacturing integration. The co-investigators of MML are Mark Cutkosky and Kos Ishii from Mechanical Engineering and Hau Lee from Industrial Engineering and Engineering Management. The laboratory has working relationships with the Graduate School of Business (GSB), Center for Design Research (CDR), and Rapid Prototyping Laboratory (RPL) at Stanford. The current MML research areas are 1) Life-Cycle Engineering Design, 2) Supply Chain Management, and 3) Agent-Based Concurrent Engineering. MML is also home of Stanford's graduate course on design for manufacturability.

NOTES:

The Industrial Designers Society of America

IDSA

Kristina Goodrich: Executive Director and CEO

Kristina Goodrich is the executive director and CEO of the Industrial Designers Society of America (IDSA) having previously served as IDSA's deputy executive director. Kristina's dedication to the Industrial Design Excellence Awards (IDEA) program led to the sponsorship of Business Week Magazine *and her efforts have attracted coverage of the industry by significant international journalistic organizations including* CNN, CNBC, Fast Company, *the* Los Angeles Times, *and the* Washington Post. *Kristina has represented the industrial design profession at congressional hearings and has been recognized as an expert by the National Endowment for the Arts. Kristina received her B.A. in Scandinavian studies and in history, and has won numerous awards for writing and editing publications from the American Society of Association Executives.*

The IDSA quarterly publication *Innovation*

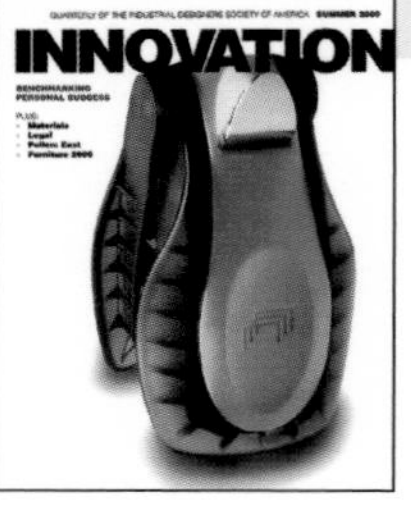

The IDSA Mission

The Industrial Designers Society of America is a professional organization serving its members in the field of industrial design. IDSA is dedicated to communicating the value of industrial design to society, business, and government. IDSA provides leadership to and promotes dialog between practice and education. As a professional association, it serves its diverse membership by recognizing excellence, promoting the exchange of information, and fostering innovation.

The IDSA Comprehensive Description of Industrial Design

Industrial design is the professional service of creating and developing concepts and specifications that optimize the function, value, and appearance of products and systems for the mutual benefit of both user and manufacturer.

Industrial designers develop these concepts and specifications through collection, analysis, and synthesis of data guided by the special requirements of the client or manufacturer. They are trained to prepare clear and concise recommendations through drawings, models, and verbal descriptions.

Industrial design services are often provided within the context of cooperative working relationships with other members of a development group. Typical groups include management, marketing, engineering, and manufacturing specialists. The industrial designer expresses concepts that embody all relevant design criteria determined by the group.

The industrial designer's unique contribution places emphasis on those aspects of the product or system that relate most directly to human characteristics, needs and interests. This contribution requires specialized understanding of visual, tactile, safety and convenience criteria, with concern for the user. Education and experience in anticipating psychological, physiological, and sociological factors that influence and are perceived by the user are essential industrial design resources.

Industrial designers also maintain a practical concern for technical processes and requirements for manufacture; marketing opportunities and economic constraints; and distribution sales and servicing processes. They work to insure that design recommendations use materials and technology effectively and comply with all legal and regulatory requirements.

In addition to supplying concepts for products and systems, industrial designers are often retained for consultation on a variety of problems that have to do with a client's image. Such assignments include product and organization identity systems, development of communication systems, interior space planning and exhibit design, advertising devices and packaging, and other related services. Their expertise is sought in a wide variety of administrative arenas to assist in developing industrial standards, regulatory guidelines, and quality control procedures to improve manufacturing operations and products.

Industrial designers, as professionals, are guided by their awareness of obligations to fulfill contractual responsibilities to clients, to protect the public safety and well-being, to respect the environment, and to observe ethical business practice.

NOTES:

DEVELOPMENT ORGANIZATIONS

Two useful resources to know about

DMI: Design Management Institute

The Design Management Institute is striving to become the international authority and champion of design management while encouraging quality management in international organizations to utilize optimum design processes. The Institute's objectives include helping design managers to become better educated leaders; collecting, organizing, and offering a variety of information and knowledge; conducting, encouraging, and funding research; and emphasizing the societal importance of design. DMI seeks to improve the public's understanding of the essence, process, and significance of design because it believes the future of design will improve the world's products and environments. The Institute was founded by Bill Hannon at the Massachusetts College of Art in Boston in 1975 and DMI became an independent nonprofit organization in 1986. In 1987, DMI established the Case Study Research Program, which is used by many prominent universities including Harvard Business School, MIT, UC Berkeley, and the London Business School. DMI members include design managers, corporate executives, independent consultants, academicians, and public sector individuals. In addition to the extensive network of professional contacts that DMI offers, DMI also sponsors conferences that present the latest developments in the design arena. The Institute also publishes an assortment of design literature including the Design Management Journal, Case Studies (which are distributed by Harvard Business School Publications), and DMI News, a bimonthly newsletter.

CDF: Corporate Design Foundation

The Corporate Design Foundation is a nonprofit organization that offers design and business education and research to those who wish to integrate these two fields into an effective and profitable relationship. Founded in 1984, CDF utilizes architecture, communication design, and product design to improve the effectiveness of businesses while also educating corporate leaders, executives, and managers about the importance of design. CDF's goals include achieving an understanding of design and identifying the role it plays in organizational prosperity, promoting innovation through education, and developing cooperative relationships between design and business schools to further their understanding of the design process. The Foundation has successfully collaborated with business school faculty to establish the first courses at schools about design and business, facilitated the development of the first interdisciplinary courses that brought together students and faculty of design, business, and engineering, and has also funded business school research, courses, and teaching materials development through design leadership grants. CDF's Board of Advisors include such prominent individuals as Jerry Hirshberg, President of Nissan Design International, and David Kelley, Principal of IDEO Product Development. Its sponsors include IBM Corporation, IDEO Product Development, Lunar Design, SoMA, Inc., and ZIBA Design.

NOTES:

A dictionary of product development terms

A

***ADD.* Analysis Driven Design.** Design methodology that utilizes analysis and test processes, both virtual and physical, throughout the product design process to drive optimal design development. Pages 044, 045.

Agency Compliance/Certification. Component or product compliance with specific safety, environmental, and/or ecological regulations resulting in official approval. Required government compliance to regulations for products in specific categories such as medical (FDA), safety (UL/CSA), etc. Page 075.

Anthropometrics. Measurements of the human body.

Appearance Design. Design of the aesthetic form, features, and style of a product.

B

Benchmark. Measurement or test used to compare the performance of a product or process against competing products or processes.

Beta. Early limited or restricted release of a product used primarily as a validation test or for evaluation.

Blue Sky. Applied to thinking without constraints regarding development issues such as cost, technology, manufacturing, etc. Used in early concept and problem-solving stages.

***BOM.* Bill of Materials.** Structured listing of all components, parts, and subassemblies used to create a product or system. Page 083.

Brainstorming. Thinking process where any and all ideas are viable, addressable, and presentable with varying degrees of criticism allowed. Page 035.

Branding. Process of developing, marketing, and selling a product and/or service using product and corporate name recognition, labeling, graphics, advertising, and experience. Pages 096, 097.

Breadboard. Rough concept mockup or schematic of a device, circuit, or product. Originally comes from electronics design.

Business Incubator. Organization that specializes in nurturing and supporting entrepreneurial start-up companies and efforts. Pages 100, 101.

C

***CAD/CADD.* Computer Aided Design/Drafting.** Using computers as tools for engineering design and drafting. Page 069.

***CAE.* Computer Aided Engineering.** Using computers to develop engineering design and analysis. Pages 024, 025.

***CAID.* Computer Aided Industrial Design.** Using computers to assist in developing the industrial design for products. Pages 048, 049.

***CAM.* Computer Aided Manufacturing.** Using computers to assist in the product manufacturing process. Page 066.

***CFD.* Computational Fluid Dynamics.** Computer-based analysis process used to predict fluid flow over a body or throughout a system. Often associated with thermal analysis.

***CNC.* Computer Numerical Control.** Process using computer controlled multi-axis machining to create precise prototypes, parts, or components. Pages 066, 067, 068.

Collaborative Equity. Creating distributed value among collaborators in an enterprise or product development. Pages 078, 079.

Collaborative Innovation. Innovation that relies on the multiple talents of an interactive group instead of an individual. Page 007.

Concurrent Engineering. Process of multiple engineering functions taking place simultaneously (or nearly so) during the development process.

Corporate Identity. Identifiable look or appearance, generally graphic, that is used throughout all products and communications within a company. Pages 088, 096, 097.

Cosmetics. External appearance or aesthetic features of a product. Sometimes deemed a derogatory label.

Critical Path. Specific path, among others, in a schedule that drives the minimum time of development completion.

Cross-Functional Team. Group of collaborating individuals whose members have varied backgrounds, capabilities, and functions.

D

Debug. Iterative process by which a product, circuit, system, process, or component is made error-free through analysis and testing. Originally comes from software engineering.

Deliverables. Agreed upon items handed over or transferred at the culmination of a phase of a project.

Design Review. Collaborative critique and discussion of design concepts, prototypes, specifications, features,

Glossary (continued)

and/or ideas. Often a specific and structured interaction at the end of each product development phase.

Design Strategy. Use of design to create a business or marketing advantage in a product or product line. Also the system of tactics to execute a product design. Page 088.

Design Vocabulary. Product appearance features that distinguish one product from another and communicate a specific style and function.

Development Enterprise. Entire system surrounding the development of a product including corporate structure, engineering, finance, human resources, research and development, manufacturing, marketing, sales, etc.

***DFA.* Design For Assembly.** Design methodologies used to optimize the ease of assembly of products.

***DFE.* Design For Environment.** Design methodologies used to reduce negative environmental impact of products and processes.

***DFM.* Design For Manufacture.** Design methodologies used to ensure that a product's or system's manufacturability is optimized.

Dimensional Performance Evaluation. Analysis of a part, component, subassembly, or product regarding its conformance to dimensional specifications. Pages 076, 077.

***DVT.* Design Validation Testing.** Testing carried out to ensure that a product design conforms to its performance specifications. Page 075.

E

***ECA, ECB.* Electronic Circuit Assembly, Electronic Circuit Board.** See PCB. Page 052.

***ECAD.* Electronic Computer Aided Design.** Using computers to develop electrical and electronic designs, schematics, systems, and components. Often associated with electronic circuit board development.

***ECO.* Engineering Change Order.** Document used to facilitate and communicate revised features or functions of a product, component, or part.

Electromechanical Design. Combination of electrical/electronic and mechanical design into a single function applied to products, parts, systems, and components.

***EMI.* Electromagnetic Interference.** Disruption of operation of an electronic device, component, product, or system caused by an electromagnetic field.

Enterprise Information Management. Comprehensive capture, organization, distribution, and maintenance of all relevant data in a development enterprise. Pages 058-063.

***EOL.* End Of Life.** Point in time when a product ceases to be manufactured or viable in the marketplace.

***ERD.* Engineering Requirements Document.** Set of specifications defining the product performance objectives in terms of engineering metrics.

***ERP.* Enterprise Resource Planning.** Process by which an enterprise effectively staffs projects, functions, and departments to meet present and future development needs.

F

***FDM.* Fused Deposition Modeling.** Rapid prototyping process using thermoplastic materials where three-dimensional physical parts are quickly created directly from CAD files. Page 066.

***FEA.* Finite Element Analysis.** Mathematical computer analysis/test process that applies various kinds of loads to a virtual model for evaluating design integrity and performance.

Focus Group. Group of individuals brought together to help explore or determine how a product might perform in the marketplace. Pages 089, 094.

Forecast-Based Demand. Using industry forecast analyses to predict the future sales of a product.

***FTP.* File Transfer Protocol.** Internet based method of transferring files from one computer to another.

G

Geometric Forms/Surfaces. Developed shapes and surfaces of a part or component that are based on simple geometry.

***GUI.* Graphical User Interface.** Pronounced "gooey." Visual product interface system that presents product operational, control, functionality, and usability features.

H

Hand-Off Management. Managing the transfer of responsibilities and information from one group to another in a project.

Glossary (continued)

Hard Tooling. Manufacturing device(s) used to create individual parts for robust, high volume mass production. Page 070.

Human Factors. Design elements, features, and issues of a product that relate to its usability and ergonomic performance. Page 024.

I

***ID.* Industrial Design.** Design discipline responsible for the form and function of a product related to its appearance, external features, human factors, ergonomics, usability, interaction, and branding. Page 118, many others.

Ideation. Iterative creative process of developing multiple ideas, concepts, and solutions. Pages 034-037.

***IDSA.* Industrial Designers Society of America.** Professional organization whose mission is to communicate the value of industrial design to society, business, and government. Page 118.

In-House. Resources available and employed within the corporate enterprise. Pages 020, 085.

Incubator. See Business Incubator. Pages 100, 101.

Interaction Design. Design methodology that focuses on the interaction and interface between users and products and the cognitive exchange of information between them.

***IT.* Information Technology.** Also known as MIS (Management Information Systems). Group assigned to manage the computer-related infrastructure and data within a corporation.

J

***JIT.* Just In Time.** Inventory method that enables delivery of parts to an assembly line as late as possible, thus reducing inventory costs and improving cash flow.

K

Knock-off. Copy or imitation of an existing product intended to capitalize on its success.

L

Lead Time. Amount of time required from the beginning of a process or task to its completion.

Life Cycle. Period of time in which a product is deemed to have usability and/or marketability. Page 071.

Logo/Logotype. Visually distinctive signature or trademark of a product or enterprise. Page 096.

M

***MCAD.* Mechanical Computer Aided Design.** CAD as applied to mechanical design and engineering. Page 024.

Mechanical Package Design. Process of designing a product's various mechanical enclosure and internal structural components, including product configuration and architecture, to meet goals of appearance, manufacturability, usability, quality, functionality, and cost. Page 028.

***MIS.* Management Information Systems.** See IT (Information Technology).

Mockup. Quickly executed physical concept model that simulates an entity's appearance, function, or usability. Pages 006, 039, 040, 068, 099.

Modular Design. Design approach involving interchangeable components for the purposes of flexibility, expandability, upgradeability, and/or increased marketable life of a product or system.

***MRD.* Marketing Requirements Document.** Specifications required to market and sell a product or service.

***MRP.* Materials Resource Planning.** Planning process system for management of product manufacturing, inventory accuracy, and distribution orders. Page 058.

N

***NPI.* New Product Introduction.** Process of preparing a company's infrastructure for the manufacturing and servicing of a newly designed product. Page 071.

***NRE.* Non-Recurring Expense/Engineering.** One-time cost(s) associated with a manufacturing process for a new design.

O

***OEM.* Original Equipment Manufacturer.** Applied to a company that manufactures a base product or subassembly that is purchased by another company for enhancement or modification and subsequent reselling under its own brand.

Operations. Corporate group responsible for the manufacturing, logistics, inventory, service, and shipping of a product.

Glossary (continued)

Organic Forms/Surfaces. Developed shapes and surfaces that are non-geometric and irregularly curved. Page 049.

***OTS.* Off-The-Shelf.** Components, parts, subassemblies, or products that are readily available without custom production.

Out-of-the-Box. Applied to thinking with perspectives and approaches different from the day-to-day norm.

Outsourcing. Using resources retained external to the corporate enterprise. Pages 020, 021, 085.

P

Para-Management. Use of a support associate to offload day-to-day routine activities of management, allowing a manager to focus on visionary and innovative leadership. Pages 016, 017.

***PCB.* Printed Circuit Board.** Circuit board assembly typically used in an electronic product. Also termed ECB and ECA. Page 052.

***PDM.* Product Data Management.** Process that continually gathers, classifies, and organizes project data on engineering modifications, maintains precise records of all relevant information, and provides immediate access to any project stage.

Perceived Value. A customer's perceived worth of a product or product feature.

Pilot Build/Run. Low quantity manufacture of a product, usually for early production evaluation.

Product Configuration. Systematic arrangement of a product's forms, features, structure, mechanical elements, and components. Pages 052, 053.

Product Configuration Design. Design of a product's configuration for optimized size, appearance, cost, features, functionality, and manufacturability. Pages 052, 053.

Product Design. Process of creating product solutions relative to usability, appearance, and mechanical structure. Often a combination of industrial design and mechanical design. Pages 028, 116, 117.

Product Development. Comprehensive process of creating a manufacturable product from concept to production to market.

Product Engineering. Engineering design and knowledge as applied to developing a product as opposed to developing a technology. Pages 028, 029.

Product Graphics. Graphics, symbols, and verbiage applied to a product's exterior to communicate its features, function, or use.

Product Innovation Content. Value and amount of innovation in the features, functionality, and performance of a product.

Product Language. Distinct product appearance features that communicate uniformity, functionality, style, quality, brand, interaction, and usability.

Productization. Process of turning a developed technology into a usable product from concept to production. Pages 028, 029.

Prototype Run. Manufacture of a small number of prototypes of a product for testing and evaluation purposes. Pages 070, 071.

Q

***QC.* Quality Control.** Management of the quality of production products as it relates to features, robustness, reliability, maintenance, and other issues affecting the performance of the product in the marketplace. Includes development and conformance to corporate quality specifications, standards, and documents.

R

***R & D.* Research and Development.** General enterprise of creating, improving, and/or enhancing existing or future products and services.

Rapid Tooling. Specialized, quickly fabricated tooling for manufacturing parts in a shorter than conventional period of time. Pages 066, 070, 071.

Regulatory Agencies. Governmental agencies that certify products for safety, materials, and manufacture. Examples are: Food and Drug Administration (FDA), Canadian Standards Association (CSA), and Underwriters Laboratories (UL).

***ROI.* Return on Investment. Return on Innovation.** Increase or decrease in either monetary or other benefit received from investing finances or resources into an enterprise or project.

Reverse Engineering. Process by which engineering design knowledge is gained by starting with a finished product or physical entity and proceeding to leverage its data content into new designs. Pages 072, 073.

***RFQ.* Request For Quotation.** Request made to an entity or organization to provide formal written estimates of cost and time to provide goods or services.

Glossary (continued)

***RPM.* Rapid Prototyping and Manufacturing.** Various techniques, processes, and materials for creating prototypes or finished products rapidly. Pages 066-071.

S

Secondary Operation. Function performed to add features to a part or product after it has been fabricated. Examples are: graphics, labels, drilled holes, finishes, etc.

Sensorize™. Simulation of an idea, concept, product, feature, or principle in a manner appropriate to the human sense(s) that should perceive it. For example, presenting an idea for audio with sound simulation instead of just describing it with words or charts. Page 006.

Setup Cost. One-time cost associated with starting production of a component or assembly.

Sketch Models. Quickly executed three-dimensional representations of a physical concept. Page 039.

***SLA.* Stereolithography Apparatus.** Rapid prototyping process and equipment using lasers and resin to rapidly produce physical parts directly from CAD files. Pages 066, 067.

***SLS.* Selective Laser Sintering.** Rapid prototyping process and equipment using plastic powder in which three-dimensional physical parts are quickly created directly from CAD files. Page 066.

Soft Tooling. Tooling, often for short runs, that is more easily and quickly created due to specialized materials and processes. Contrasted with hard tooling created from highly robust materials and processes for high volume mass production. Pages 070, 071.

Solid Modeling. Three-dimensional CAD process by which designs are developed in virtual volumetric "solid" masses. Pages 024, 025.

Structural Analysis. Process by which computer-based mathematical tools are used to test the mechanical performance of a component or product design. Pages 024, 044.

T

Technology Corridor. Geographical area where high technology companies exist, proliferate, and thrive. Page 015.

Thermal Analysis. Process by which mathematical tools are utilized to test the thermal performance of a product, component, or system. Pages 024, 025.

Thermal Design Optimization. Use of thermal analysis techniques to obtain the best balance of heat dissipation versus product features. For example, fan noise level versus product size. Page 045.

3D Printer. Equipment that creates rapid prototypes with desktop-sized machines. Page 066.

3D Scanner. Equipment that scans three-dimensional objects for importing the resultant data into a surfacing or modeling CAE or CAID program. Page 072.

***3D.* Three-Dimensional.** Term applied to virtual or real objects, models, or spaces having volume.

***TTM.* Time-to-Market.** Period of time to develop a product from concept to customer delivery.

***TT&T.* Tools, Tactics & Talent™.** Three essential components of a robust, innovative, and efficient development process.

Tooling. Device or system of devices used to give shape and form to a material during its manufacture.

Tooling Engineer. Individual responsible for the facilitation and creation of tooling. Pages 065, 074.

Turnkey. Applied to the use of contracted outside resources for executing part or all of the development or manufacture of a finished part or product with minimal supervision or management from the contracting party.

U

***UL.* Underwriters Laboratories.** Regulatory agency responsible for testing, evaluating, and certifying products for consumer safety.

***URL.* Universal Resource Locator.** World Wide Web term to describe a specific address on the Internet.

User Experience Design. The design of the aspects of a product that provide the user with more than just simple functionality, but with a positive experience associated with the product.

User Requirements. List of features or criteria that a product must possess to satisfy intended user needs.

V

Video Capture. Use of video to document ideas, processes, procedures, and events for development evaluation, documentation, or presentation. Page 060.

Videoethnography. Using video recordings to determine and analyze human behaviors. Page 089.

Glossary (continued)

Virtual Engineering. Engineering design and analysis using computer simulation. Page 025.

Virtual Market Research. Using a variety of simulation processes to evaluate market acceptance and response to product concepts and ideas. Pages 094, 095.

Virtual Prototyping. Realistic representation, visualization, and simulation of proposed product and component designs with computers as opposed to actual physical prototypes. Page 024.

Virtual Reality. Computer simulation of a real or imaginary environment allowing the user to interact with simulated surroundings.

VPD. **Virtual Product Development.** Design and development of products using computer based modeling, simulation, testing, and analysis. Page 025.

W

War Room. Physical space or room used as a work and interaction place for a project team. Page 012.

XYZ

Many who made this book possible

COMPANIES

Acuity Incorporated. River Forum Number 123, 4380 SW Macadam Avenue, Portland, OR 97201-6404. (503) 221-6995. www.acuityinc.com. Pages *xiv*, 086.

Adobe Systems Incorporated. 345 Park Avenue, San Jose, CA 95110-2704. (408) 536-6000. www.adobe.com. Colophon.

Alias|Wavefront Corporation. 210 King Street East, Toronto, Ontario, Canada M5A 1J7. (800) 447-2542. www.aw.sgi.com. Pages *xiv*, 008, 009, 038, 048.

Apple Computer, Inc. One Infinite Loop, Cupertino, CA 95014-2084. (408) 996-1010. www.apple.com. Colophon.

ARRK Product Development Group. 15201 NW Greenbrier Parkway, Building B, Beaverton, OR 97006-5771. (503) 614-2020. www.arrk.com. Page 071.

BIO/START. Third Floor, 3130 Highland Avenue, Cincinnati, OH 45219-2374. (513) 475-6610. Pages 100, 101.

Boeing Company, The. PO Box 3707, M/C 10-06, Seattle, WA 98124-2207. (206) 655-1131. www.boeing.com. Pages 062, 063.

CORBIS Corporation. 15395 SE 30th Place, Suite 300, Bellevue, WA 98007. (800) 260-0444. www.corbis.com. Page 003 (artist photos).

Ethicon Endo-Surgery, Inc. 4545 Creek Road, Cincinnati, OH 45242-2839. (513) 786-7000. www.eesonline.com. Pages *xviii*, 051.

Fender Musical Instruments Corporation. 7975 North Hayden Road, Scottsdale, AZ 85258-3246. (480) 596-9690. www.fender.com. Pages 002, 003.

Flightcom Corporation. 7340 SW Durham Road, Portland, OR 97224-7307. (800) 432-4342. www.flight-com.com. Pages 096, 097.

Industrial Light + Magic (ILM). San Rafael, CA. (415) 448-9000. www.ilm.com. Pages 042, 043.

InFocus Corporation. 27700B SW Parkway Avenue, Wilsonville, OR 97070-9215. (503) 685-8888. www.infocus.com. Pages 005, 017, 027, 045, 067, 074, 083, 092, 093.

Intrigo, Inc. 350 Conejo Ridge Avenue, Thousand Oaks, CA 91361-4928. (805) 494-1742. www.intrigo.com. Pages 078, 079.

Iomega Corporation. 1821 West Omega Way, Roy, UT 84067-3149. (801) 332-1000. www.iomega.com. Page 021.

Lexmark International, Inc. 740 West New Circle Road, Lexington, KY 40550. (859) 232-2000. www.lexmark.com. Pages 049, 053, 075, 105.

Nike, Inc. One Bowerman Drive, Beaverton, OR 97005-6453. (503) 671-6453. www.nike.com. Pages 054, 055, 090.

Nokia Corporation, Nokia Design Center. Calabasas, CA 91302. (818) 876-6000. www.nokia.com. Pages 018, 019, 107.

Rautenbach Aluminium-Technologie GmbH. Giesserweg 10, 38855 Wernigerode, Germany. www.rautenbach.de. Pages 076, 077.

Silicon Graphics, Inc. (SGI). 1600 Amphitheatre Parkway, Mountain View, CA 94043. (650) 960-1980. www.sgi.com. Pages *xiv*, 009.

Structural Dynamics Research Corporation (SDRC). 2000 Eastman Drive, Milford, OH 45150-2789. (513) 576-2400. www.sdrc.com. Pages *xiv*, 004, 100.

CONSULTANCIES

Alchemy. San Francisco, CA. www.alchemylabs.com. Pages 050, 095.

Astro Studios, Inc. 818 Emerson Street, Palo Alto, CA 94301. (650) 321-5635. www.astrostudios.com. Pages 090, 116.

bioDesign. 316 Fillmore Street, Pasadena, CA 91106-3614. (626) 744-9168. Page 037.

Designworks/USA (A BMW Subsidiary). 2201 Corporate Center Drive, Newbury Park, CA 91320-1421. (805) 499-9590. www.designworksusa.com. Pages 014, 015, 040, 073, 082.

Fitch, Inc. 10350 Olentangy River Road, PO Box 360, Worthington, OH 43085. (614) 885-3453. www.fitch.com. Pages 021, 098, 099.

frogdesign, inc. 420 Bryant Street, San Francisco, CA 94107-1303. (415) 442-4804. www.frogdesign.com. Page 50.

Function Engineering, Inc. 635 High Street, Palo Alto, CA 94301-1626. (650) 326-8834. www.function.com. Page 050.

FUSE, Inc. 1415 SE 8th Avenue, Portland, OR 97214. (503) 238-3999. www.fuseid.com Pages 005, 007, 036.

Credits (continued)

IDEO. 700 High Street, Palo Alto, CA 94301. (650) 289-3400. www.ideo.com. Pages 011, 022, 023, 033, 104.

New Product Dynamics. Portland, OR. (503) 248-0900. www.NewProductDynamics.com. Back cover.

RKS Design, Inc. 350 Canejo Ridge Avenue, Thousand Oaks, CA 91361-4928. (805) 370-1200. www.rksdesign.com. Pages 012, 013, 068, 078, 079, 091.

SoMA, Inc. 514 NW 11th Avenue, Portland, OR 97209-3227. (503) 241-1900. www.somainc.com. Pages 096, 097.

weilandesign. 211 North Malgren Avenue, San Pedro, CA 90732. (310) 548-1560. Page 039.

ZIBA Design. 334 NW 11th Avenue, Portland, OR 97209-2902. (503) 223-9606. www.ziba.com. Page 088.

ORGANIZATIONS

Corporate Design Foundation (CDF). 20 Park Plaza, Suite 321, Boston, MA 02116-4303. (617) 350-7097. www.cdf.org. Page 119.

Design Management Institute (DMI). 29 Temple Place, Boston, MA 02111-1350. (617) 338-6380. www.designmgt.org. Page 119.

Industrial Designers Society of America (IDSA). 1142 Walker Road, Great Falls, VA 22066-1836. (703) 759-0100. www.idsa.org. Page 118.

SCHOOLS

Art Center College of Design. 1700 Lida Street, Pasadena, CA 91103-1999. (626) 396-2200. www.artcenter.edu. Pages 046, 047, 089, 103, 116.

Stanford University. Stanford, CA 94305. (650) 723-2300. www.stanford.edu. Pages 035, 117.

INDIVIDUALS

Adams, James L., Ph.D. Professor Emeritus. Design Division. Stanford University. Terman 515, Stanford, CA 94305. (650) 723-1849. www.stanford.edu. Page 035.

Andersen, Mark. Assistant Chair of Product Design. Art Center College of Design. 1700 Lida Street, Pasadena, CA 91103-1999. (626) 396-2200. www.artcenter.edu. Page 116.

Anstine, Lisa. Engineering Support Specialist. InFocus Corporation. 27700B SW Parkway Avenue, Wilsonville, OR 97070-9215. (503) 685-8888. www.infocus.com. Pages 027, back cover.

Baker, Mary, Ph.D. Senior Vice President. ATA Engineering, Inc. 11995 El Camino Real, Suite 200, San Diego, CA 92130. (858) 792-3900. www.ata-engineering.com. Page 044.

Basey, Gary. Senior Product Design Engineer. InFocus Corporation. 27700B SW Parkway Avenue, Wilsonville, OR 97070-9215. (503) 685-8888. www.infocus.com. Page 045.

Beach, David. Professor-Teaching. Design Division. Stanford University. Terman 515, Stanford, CA 94305. (650) 723-3917. www.stanford.edu. Page 117.

Bennett, Katherine. Principal. Katherine Bennett Industrial Design. Santa Monica, CA. (310) 392-6081. www.kbid.net. Faculty Member at Art Center College of Design. Page 089.

Biber, Cathy, Ph.D. Senior Thermal Engineer. InFocus Corporation. 27700B SW Parkway Avenue, Wilsonville, OR 97070-9215. (503) 685-8888. www.infocus.com. Page 045.

Boyle, Dennis. Studio Leader. IDEO. 700 High Street, Palo Alto, CA 94301. (650) 289-3400. www.ideo.com. Page 033.

Bruce, Rob. Vice President and Design Director. Astro Studios, Inc. 818 Emerson Street, Palo Alto, CA 94301. (650) 321-5635. www.astrostudios.com. Pages 090, 116.

Burt, Jim. Product Design Services Manager. InFocus Corporation. 27700B SW Parkway Avenue, Wilsonville, OR 97070-9215. (503) 685-8888. www.infocus.com. Pages 066, 067.

Buxton, William. Chief Scientist. Alias|Wavefront Corporation and Silicon Graphics, Inc. (SGI). 210 King Street East, Toronto, Ontario, Canada M5A 1J7. (800) 447-2542. www.aw.sgi.com. Page 009.

Cilia, Juan. Voodoo Works (Shop) Manager. RKS Design, Inc. 350 Canejo Ridge Avenue, Thousand Oaks, CA 91361-4928. (805) 370-1200. www.rksdesign.com. Page 068.

Coleman, William. Global Marketing Manager. Alias|Wavefront Corporation. 210 King Street East, Toronto, Ontario, Canada M5A 1J7. (800) 447-2542. www.aw.sgi.com.

Connors, Thomas. President. Visual Engineering, Inc. 10105 Metropolitan Drive, Austin, TX 78758. (512) 339-8535. www.visualcnc.com.

Credits (continued)

Cooper, Aaron. Senior Industrial Designer. Nike, Inc. One Bowerman Drive, Beaverton, OR 97005-6453. (503) 671-6453. www.nike.com. Pages 054, 055.

Curtis, Alastair. Group Design Director. Nokia Corporation, Nokia Design Center. Calabasas, CA 91302. (818) 876-6000. www.nokia.com. Page 107.

Dancer, Judy. Engineering Products Manager. InFocus Corporation. 27700B SW Parkway Avenue, Wilsonville, OR 97070-9215. (503) 685-8888. www.infocus.com. Page 017.

Dauberman, Ira. Director, Global Account Manager, The Boeing Company. Structural Dynamics Research Corporation (SDRC). 600 108th Avenue NE, Suite 340, Bellevue, WA 98004. (425) 453-5232. www.sdrc.com.

Del'Ve, Robert. Senior Vice President. Designworks/USA (A BMW Subsidiary). 2201 Corporate Center Drive, Newbury Park, CA 91320-1421. (805) 499-9590. www.designworksusa.com. Page 082.

Diskin, Steve. Industrial Design Graduate Program Coordinator. Art Center College of Design. 1700 Lida Street, Pasadena, CA 91103-1999. (626) 396-2200. www.artcenter.edu. Page *xv*.

Dry, Jonathan. Lead Engineer. Mobile Phones Group, Nokia Corporation, Nokia Design Center. Calabasas, CA 91302. (818) 876-6000. www.nokia.com. Page 073.

Duncan, Michael. Northwest Regional Manager. ARRK Product Development Group. 15201 NW Greenbrier Parkway, Building B, Beaverton, OR 97006-5771. (877) 739-ARKK. www.arrk.com. Pages 066, 071.

Eustermann, John. Attorney At Law. Hutchison, Hammond & Walsh, PC. 21790 Willamette Drive, West Linn, OR 97068-3257. (503) 656-1694. Page 106.

Faste, Rolf. Product Design Program Director. Design Division. Stanford University. Terman 515, Stanford, CA 94305. (650) 723-2008. www.stanford.edu. Page 117.

Fender, Clarence Leo (Deceased). Founder. Fender Guitar Company. Fullerton, CA. Pages 002, 003.

Fisker, Henrik. President/CEO. Designworks/USA (A BMW Subsidiary). 2201 Corporate Center Drive, Newbury Park, CA 91320-1421. (805) 499-9590. www.designworksusa.com. Page 040.

Frankenstein, Carol. President. BIO/START. Third Floor, 3130 Highland Avenue, Cincinnati, OH 45219-2374. (513) 475-6610. Page 101.

Gassett, John. Senior Industrial Designer. Lexmark International, Inc. 740 West New Circle Road, Lexington, KY 40550. (859) 232-2000. www.lexmark.com. Page 049.

Goodrich, Kristina. Executive Director and CEO. Industrial Designers Society of America (IDSA). 1142 Walker Road, Great Falls, VA 22066-1836. (703) 759-0100. www.idsa.org. Page 118.

Goria, Rob. Sales Representative. Paramount Graphics, Inc. 11000 SW 11th Street, Suite 400, Beaverton, OR 97005-4126. (503) 641-7771. www.paramountgraphics.com. Page *iv*.

Haislip, Kevin. Photographer. Haisliphotography. PO Box 1862, Portland, OR 97207. (503) 254-8859.

Harker, John V. President/CEO/Chairman. InFocus Corporation. 27700B SW Parkway Avenue, Wilsonville, OR 97070-9215. (503) 685-8888. www.infocus.com. Pages 092, back cover.

Haygood, David. Vice President of Business Development. IDEO. 700 High Street, Palo Alto, CA, 94301. (650) 289-3400. www.ideo.com. Page 023.

Hearn, Garry. Industrial Design Consultant. Structural Dynamics Research Corporation (SDRC). 2000 Eastman Drive, Milford, OH 45150-2789. (513) 576-2400. www.sdrc.com.

Heermann, Thomas. Industry Marketing Manager. Design Division. Alias|Wavefront Corporation. 210 King Street East, Toronto, Ontario, Canada M5A 1J7. (800) 447-2542. www.aw.sgi.com. Page 048.

Heller, Steven. Photography. Art Center College of Design. 1700 Lida Street, Pasadena, CA 91103-1999. (626) 396-2200. www.artcenter.edu. Pages 046, 047, 103, 116 (all Art Center photos).

Holland, Gray. Vice President of Digital Design Innovation. frogdesign, inc. 420 Bryant Street, San Francisco, CA 94107-1303. (415) 442-4804. www.frogdesign.com. Alchemy. www.alchemylabs.com. Pages 038, 050, 095.

Howell, Lisa. Graphic Designer. Strategic Visuals, Inc. Vancouver, WA. (360) 571-0373. www.strategicvisuals.com. Page *iv*.

Ishii, Kosuke, Ph.D. Associate Professor. Design Division. Stanford University. Terman 515, Stanford, CA 94305. (650) 725-1840. www.stanford.edu. Page 117.

Joehnk, Terry. Principal. Contract Design. Portland, OR. (503) 287-3598. www.users.uswest.net/~jst2/. Page *xviii*.

Credits (continued)

Kalinowski, Robert, Ph.D. Editor. Hawthorne, CA. (310) 261-3137. Page *iv*.

Katz, Barry. Dean of IDEO U. IDEO. 700 High Street, Palo Alto, CA 94301. (650) 289-3400. www.ideo.com. Page 104.

Kauk, Elizabeth. Vice President/CFO. Bright Future Business Consultants. 32332 Jacklynn Court, Union City, CA 94587. (510) 489-5156. www.brightfutureconsultants.com. Page 087.

Kauk, Glen. Vice President. Bright Future Business Consultants. 32332 Jacklynn Court, Union City, CA 94587. (510) 489-5156. www.brightfutureconsultants.com. Page 087.

Kelley, David. Founder/CEO. IDEO. 700 High Street, Palo Alto, CA 94301. (650) 289-3400. www.ideo.com. Professor. Design Division. Stanford University. Stanford, CA 94305. (650) 688-3400. www.stanford.edu. Page 022.

Koshalek, Richard. President. Art Center College of Design. 1700 Lida Street, Pasadena, CA 91103-1999. (626) 396-2200. www.artcenter.edu. Pages 047, back cover.

Kovacs, Frank. Vice President of Marketing. Structural Dynamics Research Corporation (SDRC). 2000 Eastman Drive, Milford, OH 45150-2789. (513) 576-2400. www.sdrc.com.

Kraimer, Bill. Staff Industrial Designer. Ethicon Endo-Surgery, Inc. 4545 Creek Road, Cincinnati, OH 45242-2839. (513) 786-7000. www.eesonline.com. Page 051.

Lam, Jeff. Multimedia Designer. (503) 661-0525. slamjam360@earthlink.net. Page *iv*.

Lee, David. President. David Lee Design. 107 South Fair Oaks Avenue, Suite 327, Pasadena, CA 91105. (626) 449-1689. Faculty Member at Art Center College of Design.

Loewen, John. Senior Product Design Tooling Engineer. InFocus Corporation. 27700B SW Parkway Avenue, Wilsonville, OR 97070-9215. (503) 685-8888. www.infocus.com. Page 074.

Lovelady, Brett. President. Astro Studios, Inc. 818 Emerson Street, Palo Alto, CA 94301. (650) 321-5635. www.astrostudios.com. Page 090.

Luebberst, Trudi. Business Manager. Structural Dynamics Research Corporation (SDRC). 2000 Eastman Drive, Milford, OH 45150-2789. (513) 576-2400. www.sdrc.com. Page 100.

Maiers, Marty. Mechanical Engineer. PSC Scanning, Inc. Eugene, OR. www.pscnet.com. Page 069.

Marx, Brian. Finance Manager. InFocus Corporation. 27700B SW Parkway Avenue, Wilsonville, OR 97070-9215. (503) 685-8888. www.infocus.com. Page 083.

McKasson, Tom. President. Acuity Incorporated. River Forum Number 123, 4380 SW Macadam Avenue, Portland, OR 97201-6404. (503) 221-6995. www.acuityinc.com. Page 086.

McKim, Robert. Professor Emeritus. Design Division. Stanford University. Stanford, CA 94305. (650) 723-2300. www.stanford.edu.

Montgomery, Steve. Principal. bioDesign. 316 Fillmore Street, Pasadena, CA 91106-3614. (626) 744-9168. Faculty Member at Art Center College of Design. Page 037.

Mortimer, Whitney. Communications Director. IDEO. 700 High Street, Palo Alto, CA 94301. (650) 289-3400. www.ideo.com.

Mulholland, David. Senior CAD Applications Engineer. InFocus Corporation. 27700B SW Parkway Avenue, Wilsonville, OR 97070-9215. (503) 685-8888. www.infocus.com. Page 027.

Murrell, Spencer. Senior Vice President and Director. Fitch, Inc. 10350 Olentangy River Road, PO Box 360, Worthington, OH 43085. (614) 885-3453. www.fitch.com. Page 021.

Nuovo, Frank. Vice President and Chief Designer. Nokia Corporation, Nokia Design Center. Calabasas, CA 91302. (818) 876-6000. www.nokia.com. Pages 018, 019.

Olson, Jory. Principal Engineer. InFocus Corporation. 27700B SW Parkway Avenue, Wilsonville, OR 97070-9215. (503) 685-8888. www.infocus.com. Page 005.

Pelly, Charles. Founder/Consultant. Designworks/USA (A BMW Subsidiary). 2201 Corporate Center Drive, Newbury Park, CA 91320-1421. (805) 499-9590. www.designworksusa.com. Page 014.

Perine, Robert. President (designer, artist, photographer, writer). ARTRA Publishing, Inc. Encinitas, CA. (760) 436-1140. Page 002 (Leo Fender photo).

Perkins, Miles. Senior Publicist. Industrial Light + Magic (ILM). San Rafael, CA. (415) 448-2000. www.ilm.com.

Peters, Tom. Author and Business Consultant. Palo Alto, CA. Tinmouth, VT. www.tompeters.com. Back cover.

Credits (continued)

Pippen, Scottie. Professional Basketball Player. NBA. Portland Trailblazers. Pages 054, 055.

Polnoff, Craig. Mechanical HDK Engineer. Palm Computing, Inc. 5400 Bayfront Plaza, Santa Clara, CA 95052-8007. (408) 847-4897. www.palm.com. Page 065.

Rendell, George. Director, Industry Marketing. Structural Dynamics Research Corporation (SDRC). 2000 Eastman Drive, Milford, OH 45150-2789. (513) 576-2400. www.sdrc.com.

Rhein, Erika. Graphic Designer. Nokia Corporation, Nokia Design Center. Calabasas, CA 91302. (818) 876-6000. www.nokia.com.

Ringwald, Morgan. Public Relations Director. Fender Musical Instruments Corporation. 7975 North Hayden Road, Scottsdale, AZ 85258-3246. (480) 596-9690. www.fender.com.

Roberts, William W. Principal. VisuaLogos. Portland, OR. (503) 408-8277. www.visualogos.com. Page 006.

Ross, Clare. Senior Associate. Fitch, Inc. 10350 Olentangy River Road, PO Box 360, Worthington, OH 43085. (614) 885-3453. www.fitch.com.

Rudolf, Carsten. Project Manager of Computed Tomography. Rautenbach Aluminium-Technologie GmbH. Giesserweg 10, 38855 Wernigerode, Germany. www.rautenbach.de. Page 077.

Russell, Judy. Global Partner Manager, SDRC. Silicon Graphics, Inc. (SGI). 1600 Amphitheatre Parkway, Mountain View, CA 94043. (650) 960-1980. www.sgi.com.

Saddler, Dan. Photographer. Saddler Photography. Portland, OR. (503) 224-7934. www.asmporegon.com/saddler.html. Pages 024, 035, back cover (miscellaneous photos).

Salvatori, Phillip. Industrial Design Manager. InFocus Corporation. 27700B SW Parkway Avenue, Wilsonville, OR 97070-9215. (503) 685-8888. www.infocus.com. Page 093.

Sawhney, Ravi. President/CEO. RKS Design, Inc. 350 Conejo Ridge Avenue, Thousand Oaks, CA 91361-4928. (805) 370-1200. www.rksdesign.com. Pages 012, 013, 078, 079.

Schallberger, John. Principal. Advanced Media Solutions. Vancouver, WA. (800) 320-3533. www.advancedmediasolutions.com. Page *iv*.

Schoening, Mark. Co-Founder/Principal. FUSE, Inc. 1415 SE 8th Avenue, Portland, OR 97214. (503) 238-3999. Pages 007, 036.

Shannon, Jan. Consultant/Editor. InnerAction Tools for Change. Portland, OR. (503) 230-9230. Page *iv*.

Shepard, Linda Woodward. Faculty Member. Digital Media Department. Art Center College of Design. 1700 Lida Street, Pasadena, CA 91103-1999. 626.396.2200. www.artcenter.edu.

Sigafoos, Tom. SDRC Education Consortium Manager. Structural Dynamics Research Corporation (SDRC). 2000 Eastman Drive, Milford, OH 45150-2789. (513) 576-2400. www.sdrc.com. Page 004.

Skach, Christopher. Principal. Photography by Christopher. Vancouver, WA. (360) 882-0617. members.aol.com/chrisskach. Pages *xviii*, 005, 017, 027, 035, 045, 055, 067, 071, 074, 083, 086, 093, 133, back cover (miscellaneous photos).

Smith, C. Martin. Product Design Department Chair. Art Center College of Design. 1700 Lida Street, Pasadena, CA 91103-1999. (626) 396-2200. www.artcenter.edu. Pages 102, 103.

Smith, Preston CMC, Ph.D. Principal. New Product Dynamics. Portland, OR. (503) 248-0900. www.NewProductDynamics.com. Back cover.

Smith, Richard R. Author. Fullerton, CA. Page 002 (guitar, shop, and garage photos).

Spreckelmeier, Larry. Industrial Design Director. Ethicon Endo-Surgery, Inc. 4545 Creek Road, Cincinnati, OH 45242-2839. (513) 786-7000. www.eesonline.com. Page *xviii*.

Stahlman, Larry. Development Engineering Manager. Lexmark International, Inc. 740 West New Circle Road, Lexington, KY 40550. (859) 232-2000. www.lexmark.com. Page 053.

Stephens, Lori. Editor. Verbatim Publishing Services. Portland, OR. (503) 235-1687. hometown.aol.com/verbpub/home.html. Page *iv*.

Strickland, Michael. President. Strickland Design, Inc. 1233 SE Stark Street, Portland, OR 97214-1437. (503) 238-8829. www.StricklandDesign.com. Page *iv*.

Sylvester, Mark. Ambassador. Alias|Wavefront Corporation. 614 Chapala Street, Santa Barbara, CA 93101. (805) 884-7800. www.aw.sgi.com. Page 008.

Thurston, Maureen. Owner. ACCESS International. 600 Chaparral Road, Sierra Madre, CA 91024-1115. (626) 355-6295. Faculty Member at Art Center College of Design. Page 085.

Vossoughi, Sohrab. President/CEO. ZIBA Design. 334 NW 11th Avenue, Portland, OR 97209-2902. (503) 223-9606. www.ziba.com. Page 088.

Credits (continued)

Walker, Gary. Customer Service Representative. CTI Group. 2455 NW Nicolai Street, Portland, OR 97210. (503) 294-0393. www.colortechnology.com. Page *iv*.

Walsh, Alexandra. Industry Marketing Manager. Design Business Unit. Alias|Wavefront Corporation. 210 King Street East, Toronto, Ontario, Canada M5A 1J7. (800) 447-2542. www.aw.sgi.com. Page 038.

Wardle, Geoff. Corporate Relations Director. Art Center College of Design. 1700 Lida Street, Pasadena, CA 91103-1999. (626) 396-2200. www.artcenter.edu. Page 116.

Wegener, Angela. Business Development. ZIBA Design. 334 NW 11th Avenue, Portland, OR 97209-2902. (503) 223-9606. www.ziba.com.

Weiland, Herb. Principal. weilandesign. 211 North Malgren Avenue, San Pedro, CA 90732. (310) 548-1560. Page 039.

Weitzman, Maxim. Co-Founder/CEO. Intrigo, Inc. 350 Conejo Ridge Avenue, Thousand Oaks, CA 91361-4928. (805) 494-1742. www.intrigo.com. Pages 078, 079.

Williams, Robert. Senior Product Design Engineer. InFocus Corporation. 27700B SW Parkway Avenue, Wilsonville, OR 97070-9215. (503) 685-8888. www.infocus.com. Page *xviii*.

Winter, Lynn. Marketing. IDEO. 700 High Street, Palo Alto, CA 94301. (650) 289-3400. www.ideo.com.

Wittenbrock, Stevan. President. SoMA, Inc. 514 NW 11th Avenue, Portland, OR 97209-3227. (503) 241-1900. www.somainc.com. Pages 096, 097.

Wood, Chip. Vice President. RKS Design, Inc. 350 Conejo Ridge Avenue, Thousand Oaks, CA 91361-4928. (805) 370-1200. www.rksdesign.com. Page 091.

Yasukochi, Darren. Photographer. Designworks/USA (A BMW Subsidiary). 2201 Corporate Center Drive, Newbury Park, CA 91320-1421. (805) 499-9590. www.designworksusa.com. Pages 040, 073 (miscellaneous photos).

Zargarpour, Habib. Associate Visual Effects Supervisor. Industrial Light + Magic (ILM). San Rafael, CA. (415) 448-9000. www.ilm.com. Page 043.

PHOTOGRAPHERS (photo copyrights per photographer)

Cara, Roberto. Page 011 (group photos).

Casey, Sean. Page 043 (office photos).

Derek, Kent. Page 092 (John Harker).

English, Rick. Page 011 (airplane photo), page 033 (Tech Box).

Heller, Steven. See Individual Credits.

IDEO. Page 033 (product photos).

Industrial Light + Magic (ILM). Page 042 (photos courtesy of ILM).

Moeder, Steve. Page 033 (Dennis Boyle), page 104 (team photos).

Panelli, Brett. Page 054 (Pippen/Cooper photos).

Saddler, Dan. See Individual Credits.

Skach, Christopher. See Individual Credits.

Stanford University Press Service. Page 104 (Barry Katz).

Tuschman, Mark. Page 022 (David Kelley).

Yasukochi, Darren. See Individual Credits.

Zargarpour, Habib. Page 043 (sketches and product photos).

USING THE BAR CODES

A unique linking mechanism with a multitude of useful possibilities

As an additional presentation and teaching tool, a bar code has been printed on each of the book pages along with the page number above it. Each bar code can be used as a reference link to third party imagery, presentations, reports, data, or lesson plans for that page. By using technology such as digital cameras, scanners, the Internet, and various computer applications, one can create and compose custom supplemental material to support, expand, or enhance a principle or example in the book.

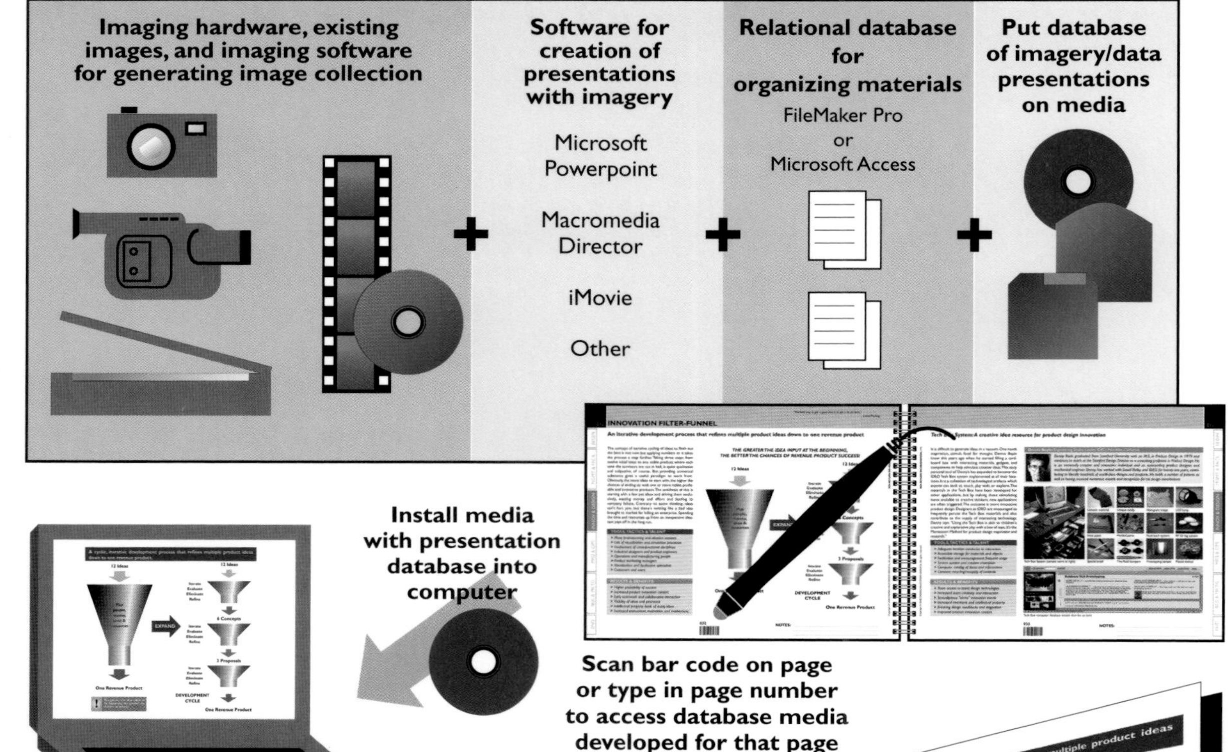

TOOLS, TACTICS & TALENT

- > *Existing imagery for presentations*
- > *Variety of image capture/generation tools*
- > *Image processing tools and tactics*
- > *Relational database for imagery files*
- > *Media copying capability*
- > *Desktop or laptop computer*
- > *Optional bar code reader linked to computer*
- > *Data/video projection system*

RESULTS & BENEFITS

- > *Expansion and enhancement of book content*
- > *Customized training and teaching materials*
- > *Self-study and homework opportunities*
- > *Remote/distance teaching and training possibilities*
- > *Visual presentation of additional examples*
- > *Design and development promotion options*

View presentation media on computer screen for selected pages

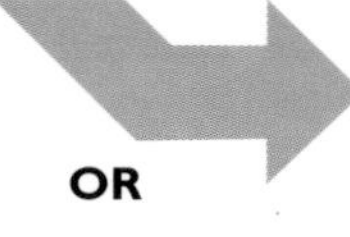

OR

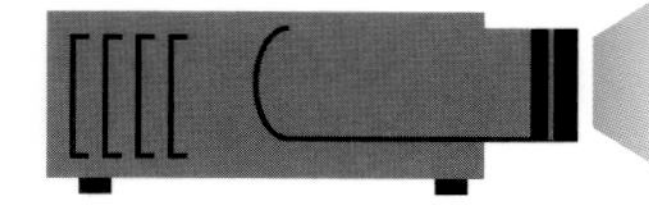

Data/video projection of presentation media for selected pages

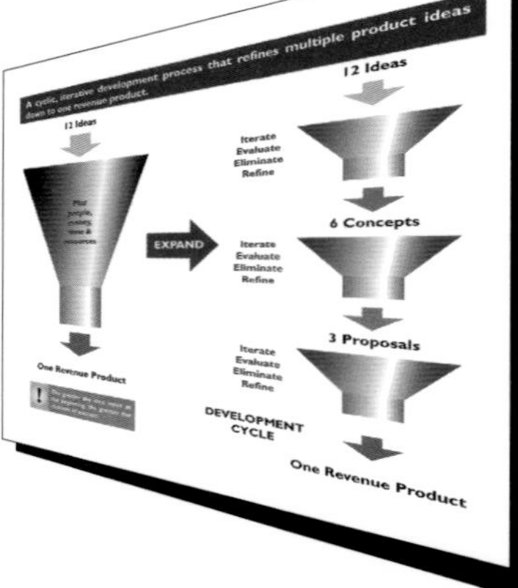

NOTES:

ABOUT THE AUTHOR

Bill Dresselhaus has spent the past thirty years in industrial design, product engineering, and product development in both corporate and consultant venues. He works between the disciplines of product engineering and industrial design. Starting his career as a chemical process engineer, he soon moved into product design and development, then to industrial design and program management, with his current emphasis on innovation facilitation. He has designed, facilitated, and/or managed several award-winning, programs and products in the high-technology marketplace from computers to projectors to chemical instruments. Bill has participated in product design, mechanical package design, industrial design, and product engineering, as well as managed all of these functions. His name is on several patents as well as on a number of international product design awards.

Bill was the first in-house product/industrial designer at Apple Computer in 1979 and was also the first such designer at InFocus Corporation in 1994. In both cases he led the product design team for each company's new ground-breaking, high-technology product—the "Lisa" at Apple and the LP210 at InFocus, both of which were revolutionary products for their times. He is featured several places as an early design innovator in the 1997 book by Paul Kunkel, *AppleDesign: The Work of the Apple Industrial Design Group*, a twenty-year history of industrial design at Apple Computer.

Bill has his own consulting practice in industrial design and product development management. He also conducts seminars and training in product development innovation and teaches industrial design, design management, physical science, and other subjects at a variety of schools and colleges. His passion is the use of visualization and simulation to enhance innovation and education.

Bill has an M.S. in product design from Stanford University (1974) and an M.S. and B.S. in chemical engineering from Iowa State and Nebraska respectively (1969/67). He has taken graduate courses and training over the past twenty-five years in industrial design at such schools as California State University at Long Beach and San Jose State University. His most recent training was in 1998 as the first pilot participant in the new Executive Special Studies Industrial Design Graduate Program at Art Center College of Design in Pasadena, California.

Things you might want to remember

Notes (continued)

Where to find it

This book is indexed by utilizing a combination of the Table of Contents, Glossary, and Credits sections. Please use these for locating words, terms, topics, people, and organizations.

COLOPHON

This book was created with Adobe Illustrator 8.0, Adobe Photoshop 5.5, and Adobe InDesign 1.5 on an Apple Macintosh 400 MHz G4 computer with OS9. Desktop page proofing was done on an Epson 1160 inkjet printer. Bar codes were created in Bar Code Pro by SNX and Adobe Illustrator. The font family is Gill Sans from Adobe Systems.

Prototype books were printed on a Chromapress digital press at Color Technology, Inc. (CTI) and coil-bound at Kinko's. Final production was done by Paramount Graphics, Inc. as follows. Pre-press was direct-to-plate using the Prinergy PDF Workflow and output to a Creo AL Trendsetter at 175 lines per inch. Printing was on 28" and 40" Komori Lithrone six-color presses with aqueous coaters using four-color process plus one spot color with soy-based inks from Ink Systems, Inc. Cover stock is 100# Luna Matte Cover and text stock is 97# Luna Matte Return Card. Wire-O binding was done by Rose City Bindery.

Many of the original computer-generated images in this book were created with Alias|Wavefront Studio, Studio Paint, and Maya CAID/SFX software or via SDRC I-Deas, Imageware, and other SDRC CAE software, generally on Silicon Graphics, Inc. (SGI) computers. Several images were created on Macintosh computers with Adobe Illustrator and/or processed with Adobe Photoshop. A few images were created on Wintel PC systems with Microsoft Windows NT.

Of course, talented people created many of the original images in this book with pencils, pens, paper, cardboard, foam, plastic, clay, hand tools, and power tools.

Have Fun,
Make Money,
Change the World!